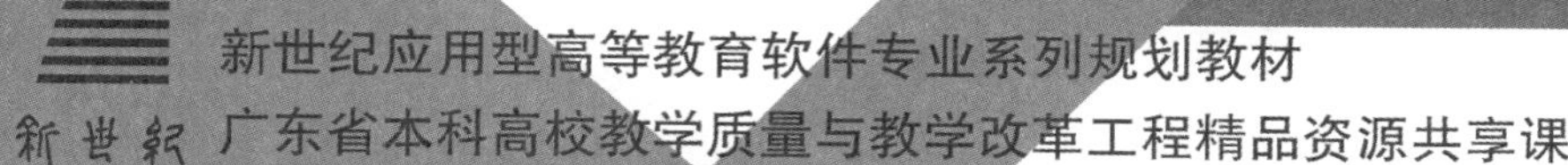

新世纪 新世纪应用型高等教育软件专业系列规划教材

广东省本科高校教学质量与教学改革工程精品资源共享课

企业级Java Web编程技术
——Servlet & JSP实验指导教程

（第三版）

QIYEJI JAVA WEB BIANCHENG JISHU
—SERVLET & JSP SHIYAN ZHIDAO JIAOCHENG

主　编　张　屹　吴向荣
副主编　谭翔纬

大连理工大学出版社

图书在版编目(CIP)数据

企业级 Java Web 编程技术. Servlet & JSP 实验指导教程 / 张屹，吴向荣主编. -- 3 版. -- 大连 ：大连理工大学出版社，2021.8

新世纪应用型高等教育软件专业系列规划教材

ISBN 978-7-5685-2917-4

Ⅰ. ①企… Ⅱ. ①张… ②吴… Ⅲ. ①JAVA 语言—程序设计—高等学校—教材 Ⅳ. ①TP312.8

中国版本图书馆 CIP 数据核字(2021)第 000586 号

大连理工大学出版社出版

地址:大连市软件园路 80 号　邮政编码:116023

发行:0411-84708842　邮购:0411-84708943　传真:0411-84701466

E-mail:dutp@dutp.cn　URL:http://dutp.dlut.edu.cn

大连永盛印业有限公司印刷　　大连理工大学出版社发行

幅面尺寸:185mm×260mm　印张:13.75　字数:318 千字

2013 年 8 月第 1 版　2021 年 8 月第 3 版

2021 年 8 月第 1 次印刷

责任编辑:孙兴乐　责任校对:王晓彤

封面设计:对岸书影

ISBN 978-7-5685-2917-4　定　价:39.80 元

前言 Preface

《企业级 Java Web 编程技术——Servlet & JSP 实验指导教程》(第三版)是新世纪应用型高等教育教材编审委员会组编的软件专业系列规划教材之一。

本教材是与《企业级 Java Web 编程技术——Servlet & JSP》(第三版)配套的实验指导教程,共有 16 个实验。每个实验包括实验目的、实验环境、实验知识背景、实验内容与步骤、实验总结和课后思考题六部分内容。"实验目的"说明了实验的主要内容和要求,通常有掌握、熟悉、了解等不同层次要求;"实验环境"说明了实验用到的软件环境要求;"实验知识背景"列举了实验相关的主要知识点;"实验内容与步骤"给出了实验的具体内容和操作步骤,题型以程序设计题为主;"实验总结"是对实验的重要性、重点、难点和实验内容进行说明和总结;"课后思考题"是对实验涉及的知识点和原理内容进行思考和探究。为了适应 Web 应用技术要求,本版教材使用了 MyEclipse 2014 插件、Weblogic 12 融合中间件等较为先进的软件平台环境,教材中所有实验均在该环境下运行并验证通过。

关于如何使用本教材以及如何分配实验课时,编者给出以下几点建议:

1.关于实验课时:本教材共有 16 个实验,以广州软件学院为例,每个学期有 17 个教学周,可以每周完成一个实验,也可以根据学校实际情况进行增加或删减。

2.关于实验环境:除了要求安装 JDK,并进行适当配置外,还推荐安装 MyEclipse 2014 插件开发平台、Weblogic 12 容器服务器和 MySQL 数据库等流行的 Java EE 开发软件,有助于提高编程效率。

3.关于课前预习:读者在上机实验前,应熟悉并掌握"实验知识背景"部分所列举的原理内容,这是实验的前提。如果达不到这一要求,应结合教材及时复习相关知识点。

4.关于课后思考题:为了对相关原理进行验证和总结,加深对主要知识点的巩固和探究,每个实验均提供课后思考题,旨在准确掌握主要知识点。

5.关于上机实践时出现的问题:对于初次接触 Java Web 编程实验的读者,上机实践时出现问题是很正常的现象。面对问题的正确态度是正视它、解决它,调试程序也是程序设计员的一种能力,这种能力需要在上机实践中提升。Java EE 中涉及的类和接口很多,我们不可能也没必要全部记住它们,但是可以通过 Java EE API 文档和 MyEclipse 的提示信息快速找到所需内容,这是程序设计员的一项基本功,应予以足够重视。

6.关于编程能力的培养:对于大多数读者来说,看懂代码比自己动手编程来得容易,这是实际操作能力不够强的表现。要想提高编程能力,最重要的是要有清晰的编程思路,并进行一定代码量的编程训练。阅读、分析他人代码,理解其编程思路是一种非常有效的途径,在此基础上模仿、改进、创新,不断积累经验和学习编程方法,才能提高编程水平。本教材有足够数量的例题,涵盖了 Java Web 应用编程的全部知识点和原理内容,能保证一定的编程训练量。

本教材由广州软件学院张屹、吴向荣任主编,广州软件学院谭翔纬任副主编。具体编写分工如下:实验 1～实验 8 由吴向荣编写,实验 9～实验 16 由谭翔纬编写。全书由张屹、吴向荣统稿并定稿。

在编写本教材的过程中,编者参考、引用和改编了国内外出版物中的相关资料以及网络资源,在此表示深深的谢意!相关著作权人看到本教材后,请与出版社联系,出版社将按照相关法律的规定支付稿酬。

鉴于我们的经验和水平,书中难免有不足之处,恳请读者批评指正,以便我们进一步修改完善。

编　者

2021 年 8 月

所有意见和建议请发往:dutpbk@163.com
欢迎访问高教数字化服务平台:http://hep.dutpbook.com
联系电话:0411-84708445　84708462

目录 Contents

实验 1 Java Web 编程环境安装配置及使用

1.1 实验目的

1.掌握 JDK 的下载与安装
2.掌握 WebLogic 12.x 的下载、安装与配置
3.掌握在 MyEclipse 中配置 WebLogic 12.x 容器
4.掌握在 Weblogic 12.x 中配置 JDK 运行环境
5.初步掌握 Java Web 项目创建、发布和运行的基本步骤
6.能够参照书中例子,编写简单的 Java Web 程序,并运行

1.2 实验环境

1.MyEclipse 插件平台
2.WebLogic(或 Tomcat)容器

1.3 实验知识背景

1.3.1 典型 Web 应用程序的结构

一个应用程序的大小和复杂性与应用程序的设计级数密切相关,程序越复杂,设计层数就越多。将应用程序合理地分成多个层,程序复杂性的控制就变得简单多了,这样开发人员就可以集中解决每个单独的层,也可以把工作切分后交给各自独立的开发小组,大大降低了开发的难度。

这里详细介绍最常用的三层结构：表示逻辑、业务逻辑、数据存取逻辑。如图 1-1 所示。

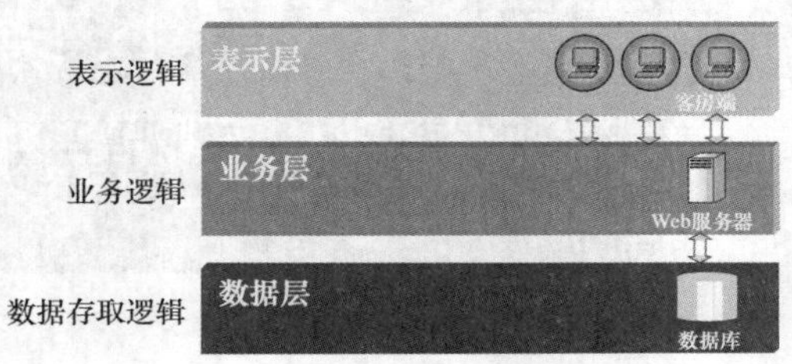

图 1-1 常用的三层结构

图 1-1 可以很清晰地看到三层结构是如何互相协作的：

(1)表示层通过一种或多种方法和使用者联系，并且负责提交和呈现数据。

(2)业务层包含应用程序在数据运算中使用的各种事务规则和操作方法。当表示层提交一个数据的时候，业务逻辑部分就对这些数据进行相应的处理，使这些原始数据变得有意义，遵循事务规则运作。

(3)数据层在应用程序中是最底层，负责保存和提取数据。

1.3.2 Web 开发过程

要了解 Web 的开发过程，我们必须先清楚它的目录结构。一个 Web 应用程序基本包括：HTML 文件、图片、Servlet、JSP 页面、JavaBean、Jar 文件、Applet、标签文件、标签库描述符文件和部署描述器等。如图 1-2 所示。

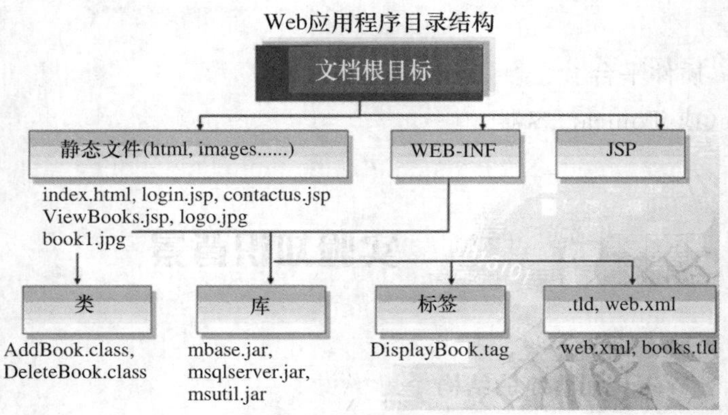

图 1-2 Web 应用程序基本目录结构

Web 程序开发过程主要包括：设计目录结构，编写 Web 应用程序代码，编写部署描述符，编译代码，将 Web 应用程序打包，部署 Web 应用程序，执行 Web 应用程序。

1.4 实验内容与步骤

1.JDK 的下载与安装

(1)进入 http://java.sun.com/javase/downloads/index.jsp 页面下载最新的安装文件。

(2)双击下载的.exe 文件,安装 JDK 到 C:\JAVA 目录下。

2.WebLogic 12.x 服务器的安装

双击从官网上下载的 WebLogic 12.x 安装文件,按安装向导一步步安装。

3.配置系统环境变量

打开系统变量对话框,以给定值设置以下各环境变量:

变量名:JAVA_HOME　变量值:C:\JAVA\jdk1.7.0

变量名:PATH　变量值:%JAVA_HOME%\bin

变量名:CLASSPATH　变量值:.;%JAVA_HOME%\lib\dt.jar;%JAVA_HOME%\lib\tools.jar

4.WebLogic 12.x 服务器的配置

(1)运行【开始】|【程序】|【Oracle】|【OracleHome】|【WebLogic Server 12c (12.1.3)】|【Tools】|【Configuration Wiazrd】,然后依次按图 1-3、图 1-4 所示步骤进行设置,选择创建一个新的 WebLogic domain。

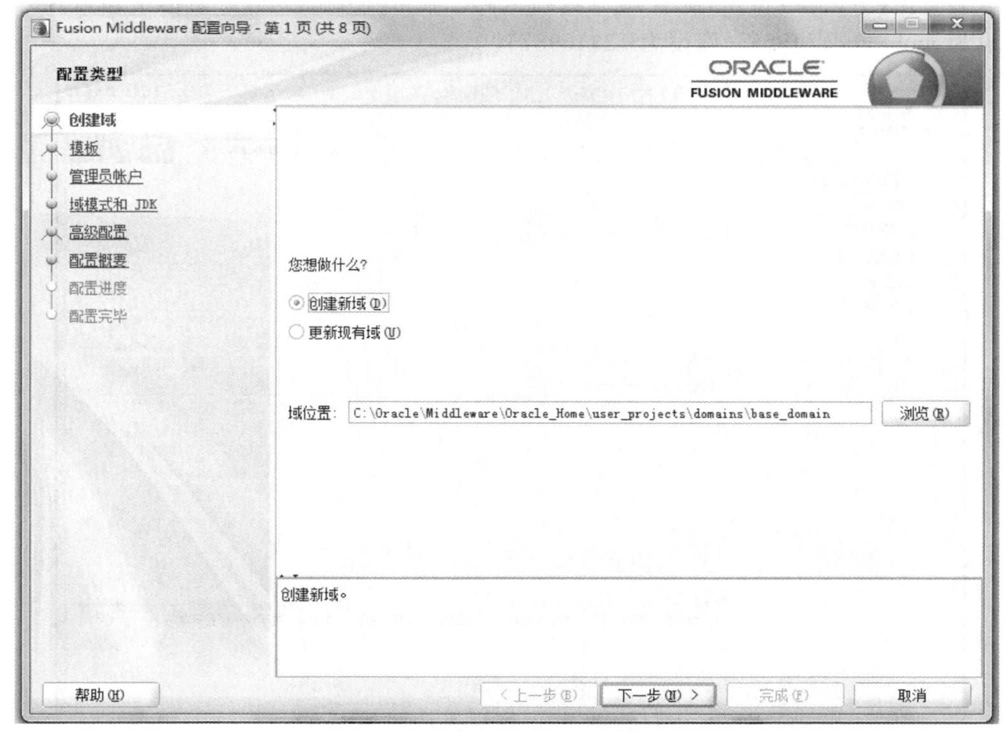

图 1-3　创建 WebLogic domain 步骤 1

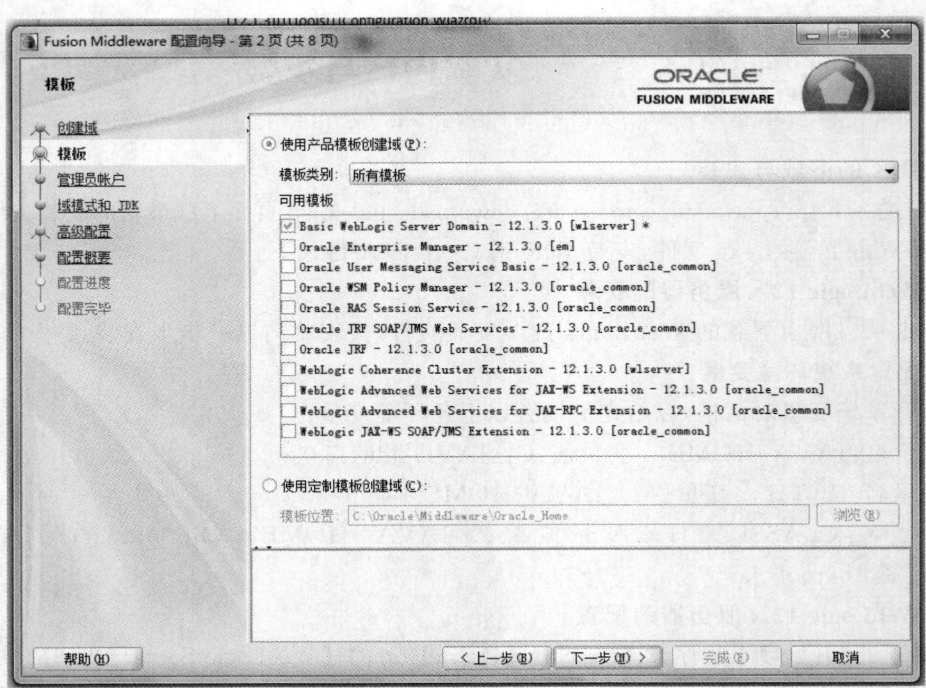

图 1-4　创建 WebLogic domain 步骤 2

(2)单击【下一步】按钮,配置用户名和密码。在如图 1-5 所示对话框中输入用户名(假设为 weblogic)和密码(假设为 12346987)。

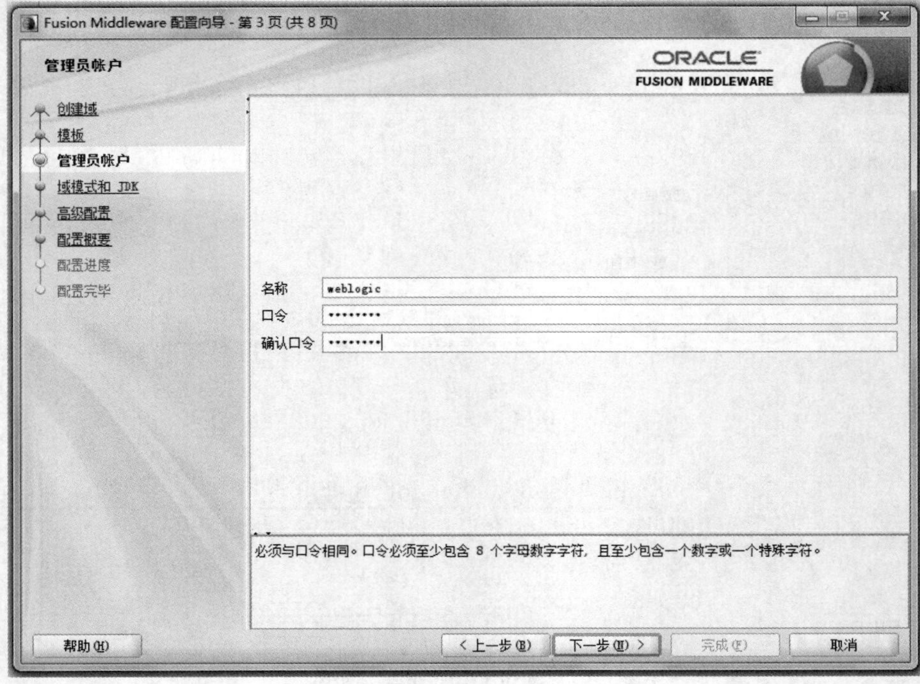

图 1-5　创建 WebLogic domain 步骤 3

(3)单击【下一步】按钮,选择开发模式和 JDK,如图 1-6 所示。

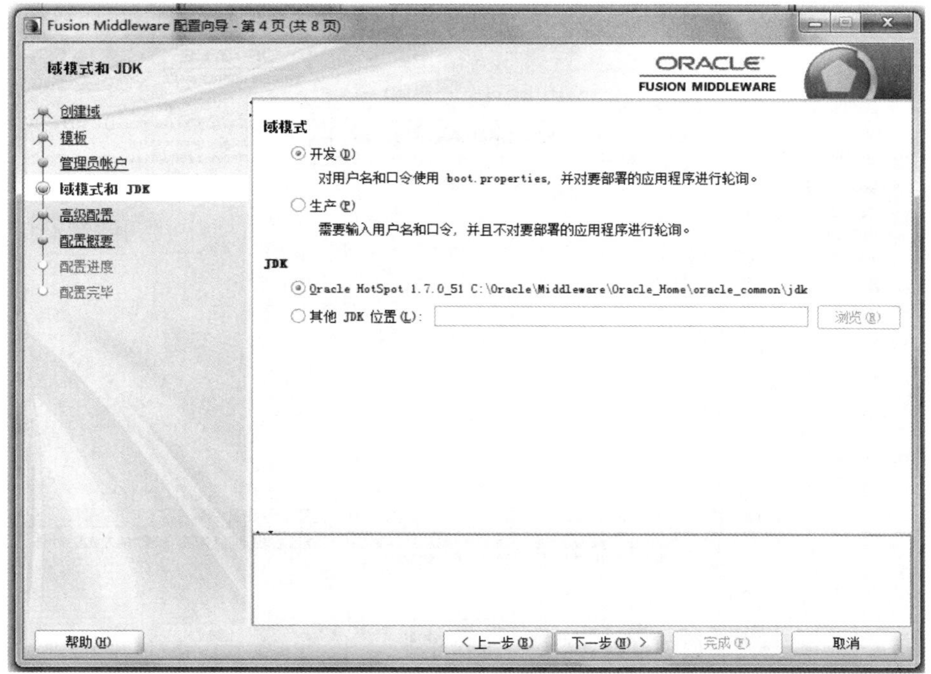

图 1-6 创建 WebLogic domain 步骤 4

(4)单击【下一步】按钮,如图 1-7 所示。

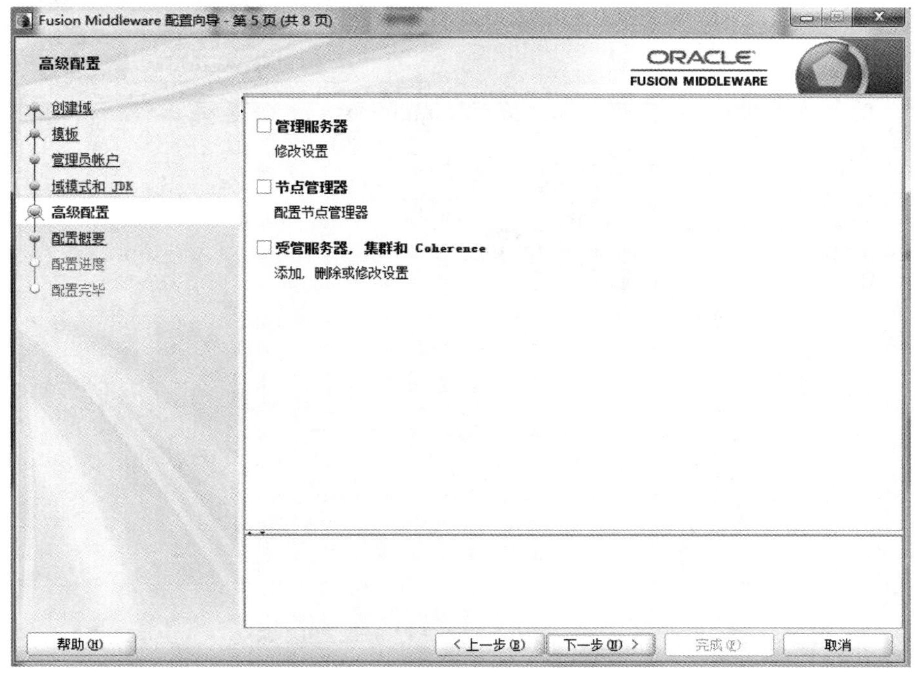

图 1-7 创建 WebLogic domain 步骤 5

(5)单击【下一步】按钮,配置 Administration Server,如图 1-8 所示。

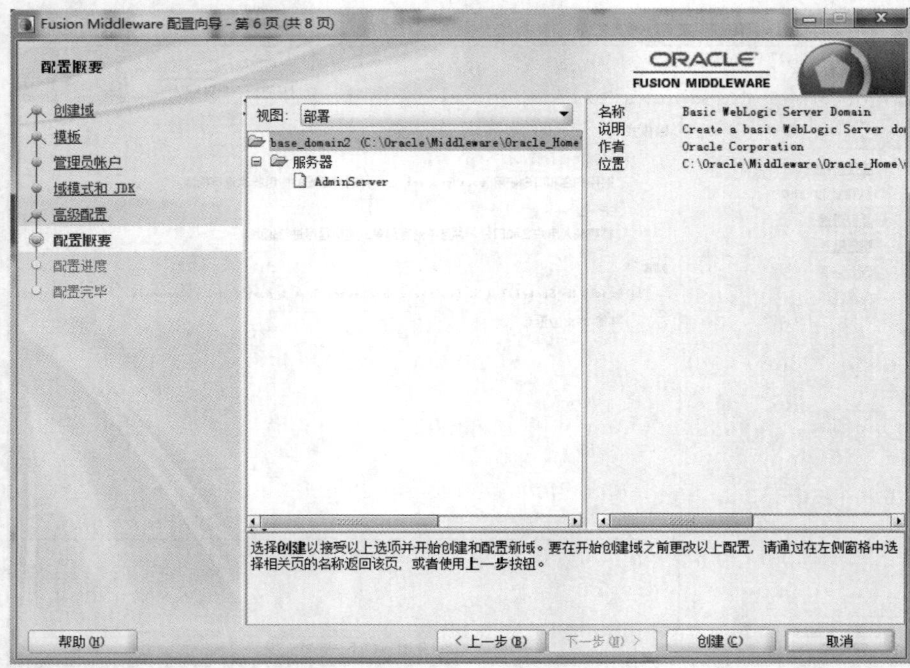

图 1-8 创建 WebLogic domain 步骤 6

(6)单击【创建】按钮,进入配置进度提示界面,如图 1-9 所示。

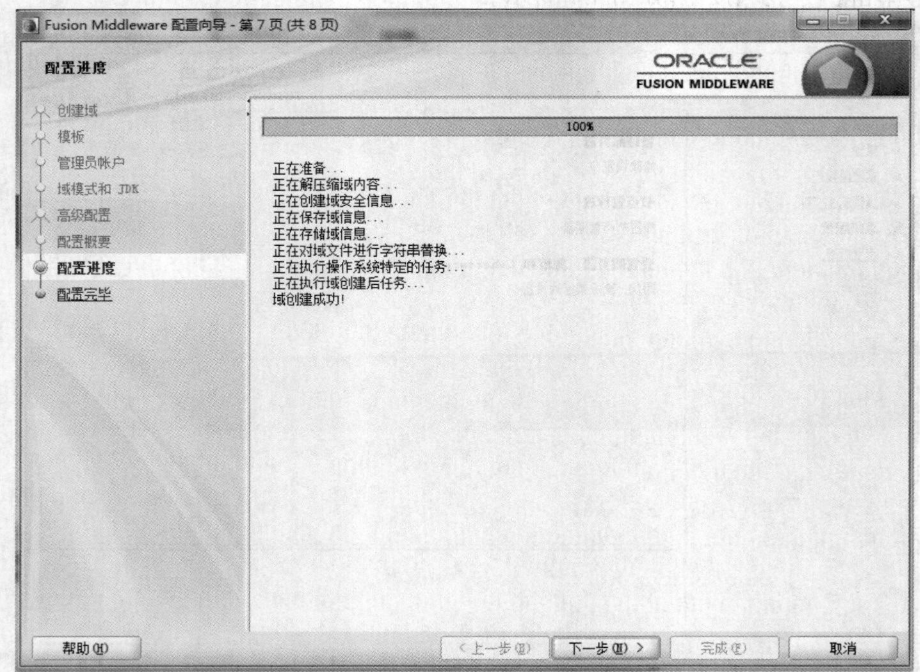

图 1-9 创建 WebLogic domain 步骤 7

(7)单击【下一步】按钮,进入配置成功提示界面,如图 1-10 所示。

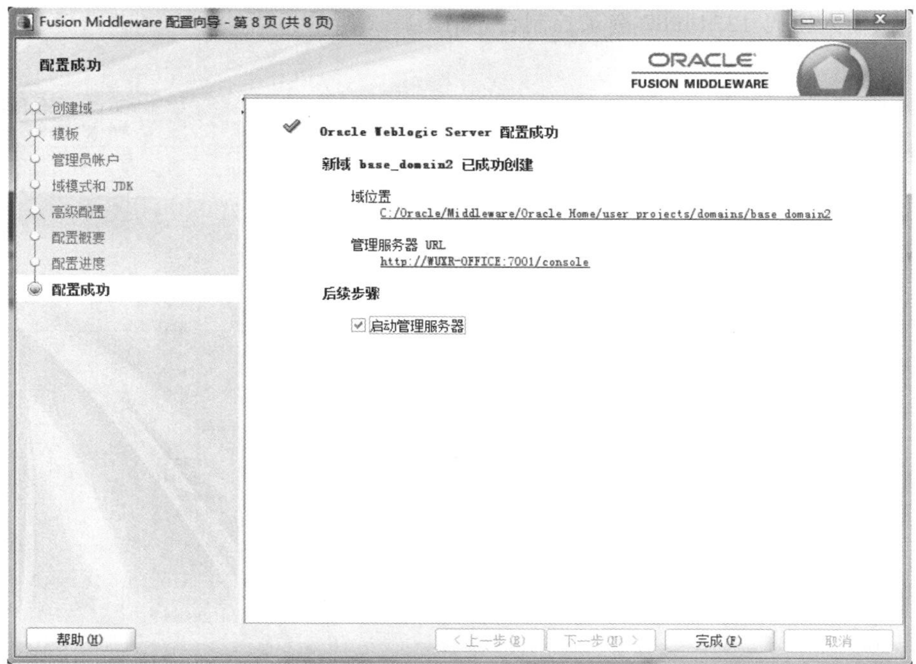

图 1-10　创建 WebLogic domain 步骤 8

(8)单击【完成】按钮,进入 Weblogic 启动界面,如图 1-11 所示(也可以进入相应安装路径使用 startWebLogic.cmd 命令启动)。

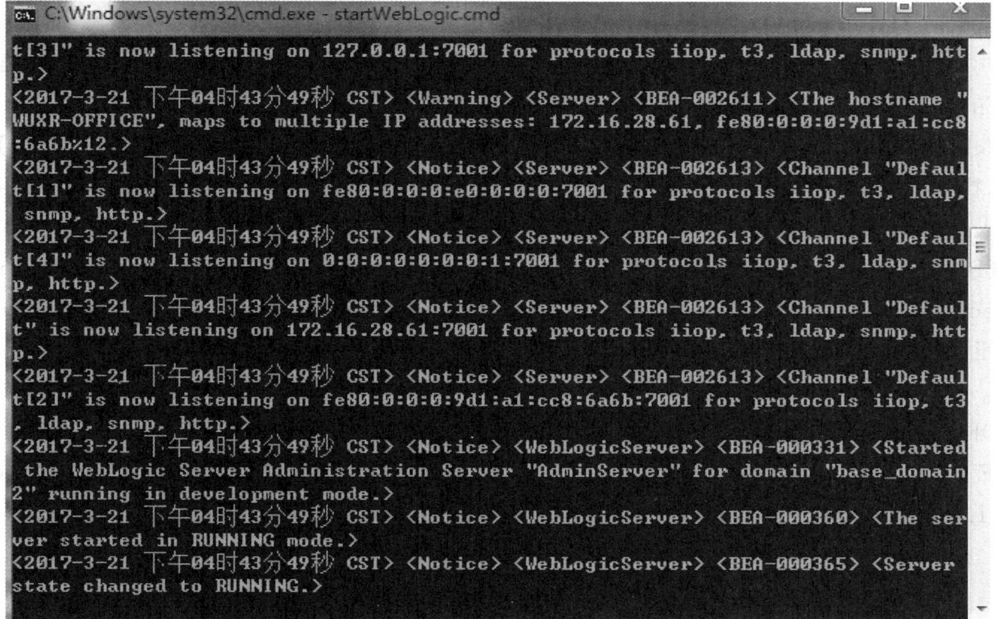

图 1-11　创建 WebLogic domain 步骤 9

(9)打开 Internet 浏览器,在地址栏输入 http://localhost:7001/console,输入用户名 weblogic,输入密码 12346987 登录控制台,如图 1-12 所示。

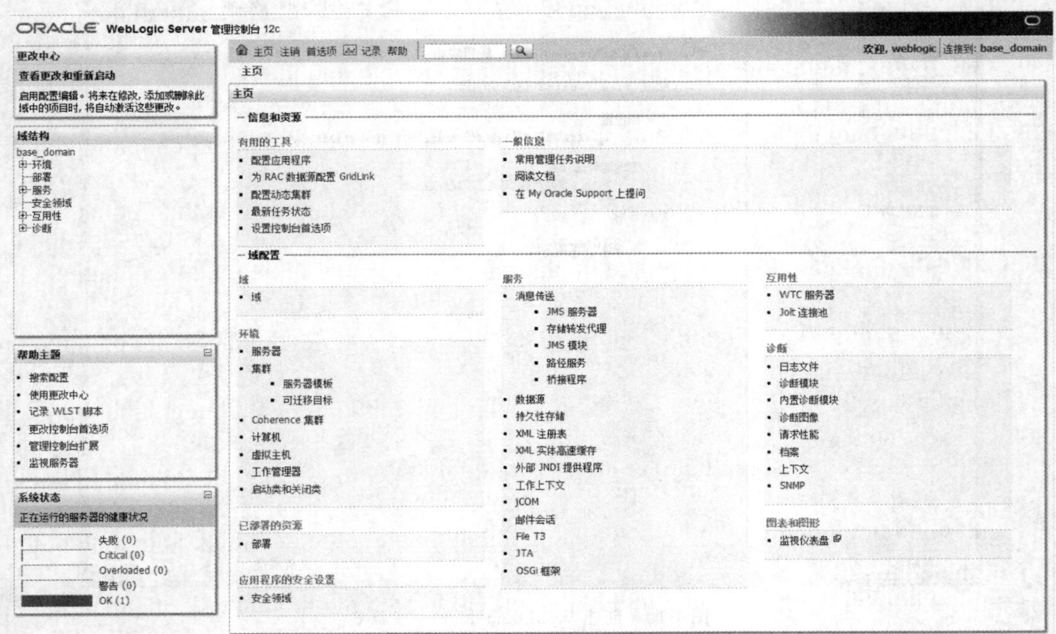

图 1-12　WebLogic 控制台

5.在 MyEclipse 中配置 WebLogic 12.x 容器

(1)在 MyEclipse 中设置 WebLogic 12.x 服务器,打开 MyEclipse 平台,单击菜单【Window】|【Preferences】,进入设置界面,鼠标单击【Servers】,展开【WebLogic】|【WebLogic 12.x】,在“WebLogic 12.x”界面进行 Weblogic server 设置。按下【Browse】按钮,找到 WebLogic 12.x 安装目录,例如 C:\Oracle\Middleware,单击【OK】按钮,在相应输入框中输入上文创建的 Administration username 和 Administration password;再次按下【Browse】按钮,找到上文创建的 base_domain 目录,例如 C:\Oracle\Middleware\Oracle_Home\user_projects\domains\base_domain,单击【OK】按钮选中。单击【Apply】按钮,然后单击【OK】按钮,如图 1-13 所示。

(2)设置 WebLogic 12.x 所使用的 JDK,单击展开【WebLogic 12.x】菜单,选择【JDK】,在右边出现的“JDK”界面上添加 WebLogic JDK 设置。单击【Add】按钮,输入 JRE name,然后单击【Browse】按钮,找到 JDK 安装路径,例如,C:\Program Files\Java\jdk1.7.0_45,单击【Apply】|【OK】按钮,完成 WebLogic Add JVM 操作。如图 1-14 所示。

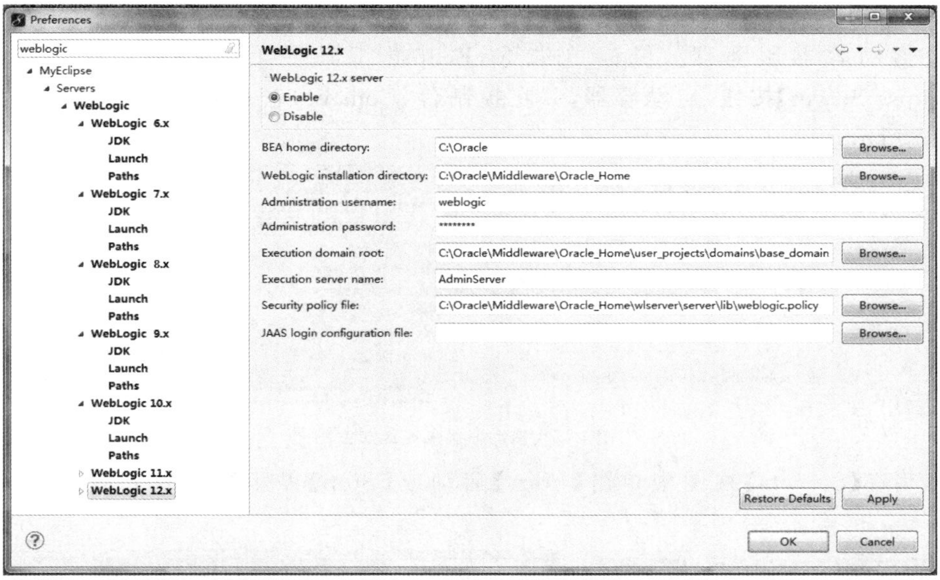

图 1-13　配置 WebLogic 12.x 容器

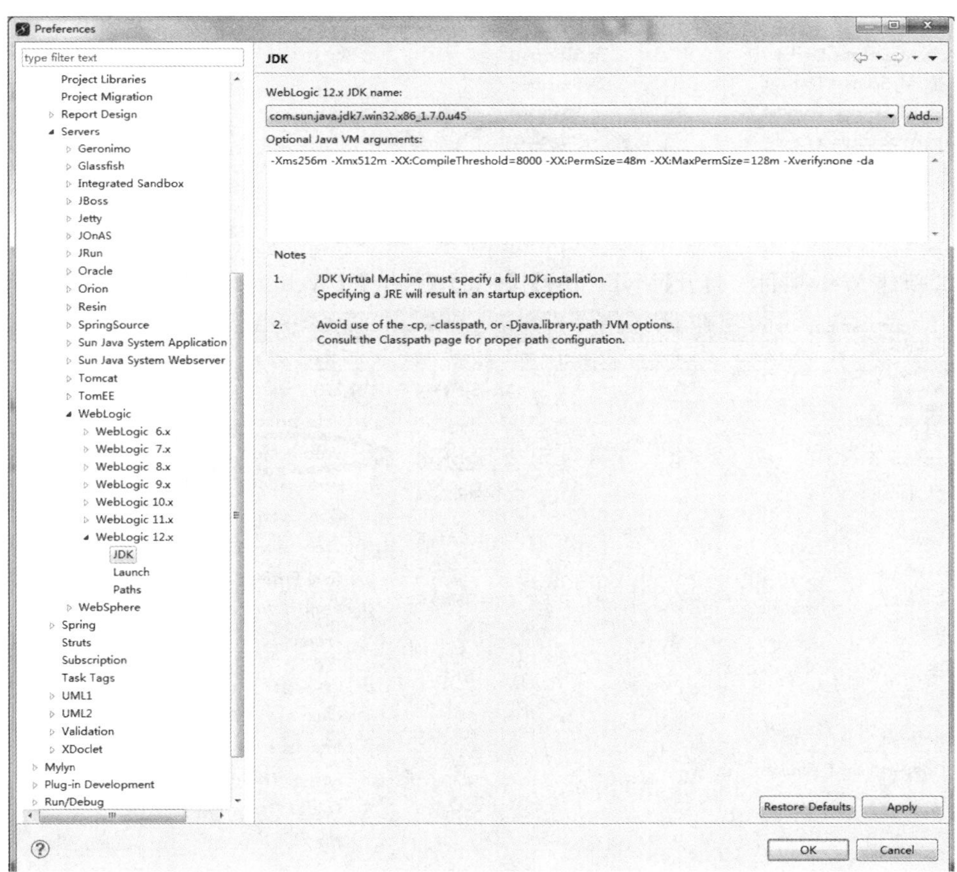

图 1-14　添加 Weblogic JDK

(3)启动 WebLogic 12.x 服务器,然后运行下列客户端程序,查看运行结果。

①启动容器服务器。鼠标单击 MyEclipse 工具栏上的【Run | Stop | Restart MyEclipse Server】按钮,启动容器,并在控制台(Console)窗口观察容器启动信息,如图 1-15 所示。

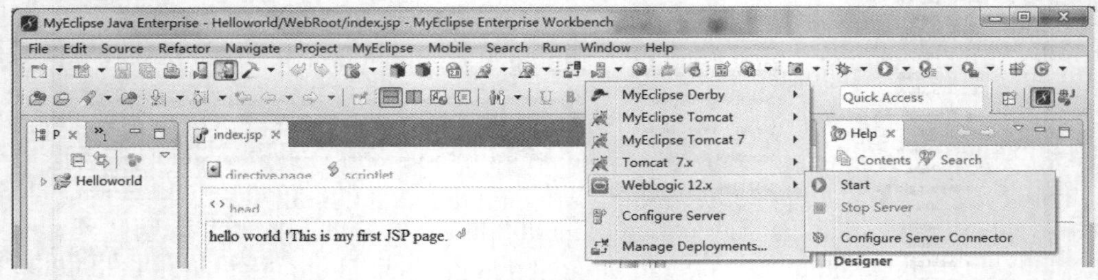

图 1-15　启动容器服务器方法 1

或者在【Servers】视图中单击【Start】启动或【Stop】停止容器服务器。如图 1-16 所示。

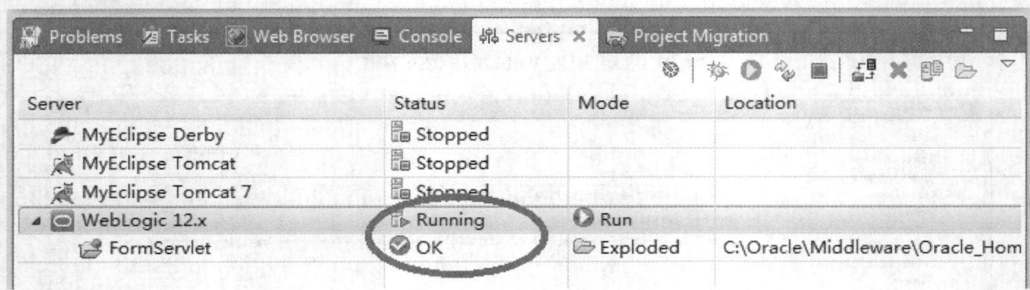

图 1-16　启动容器服务器方法 2

②新建 Web 项目。打开【MyEclipse】|【File】|【New】|【Web Project】。如图 1-17 所示。

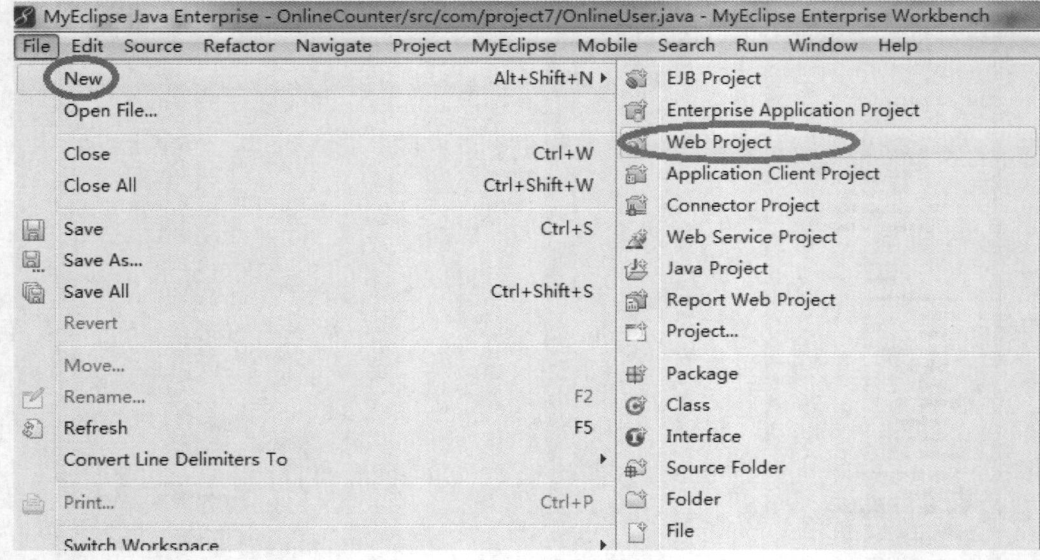

图 1-17　新建项目示意图

③填写项目信息。如图 1-18 所示。

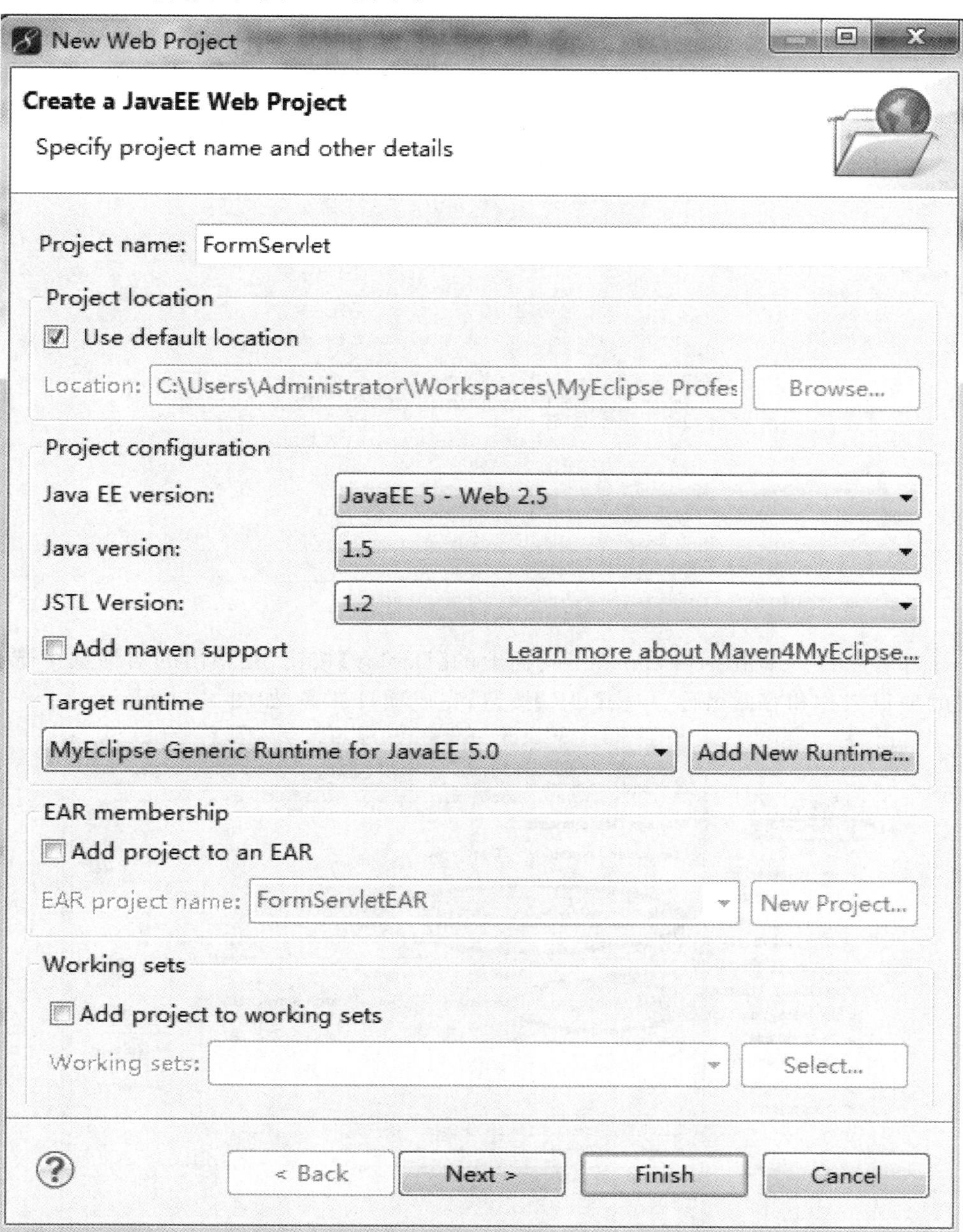

图 1-18　填写项目信息示意图

④编写代码。如图 1-19 所示。

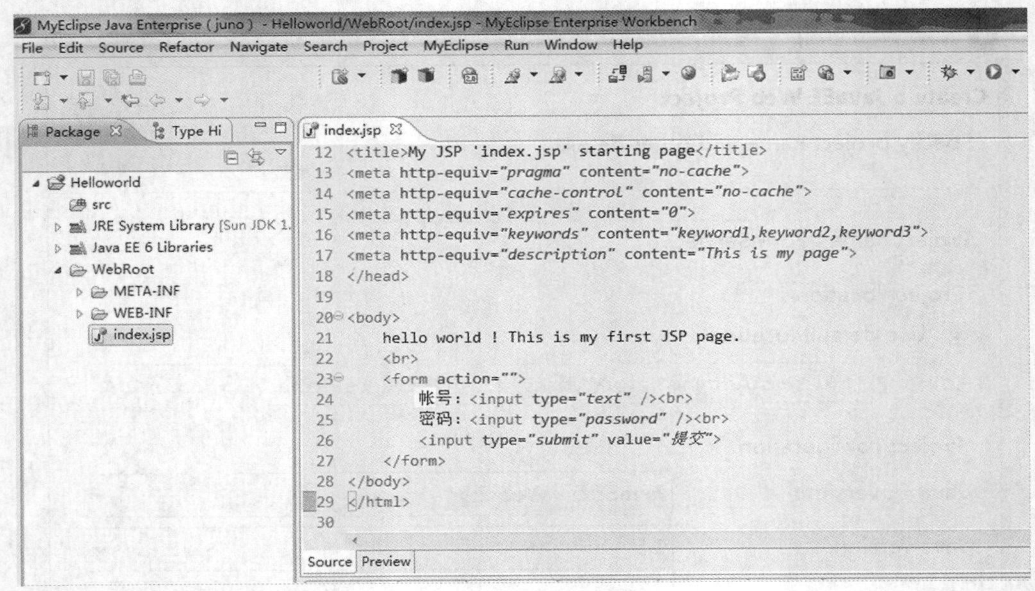

图 1-19　编写代码示意图

⑤部署项目。单击 MyEclipse 工具栏上的【Deploy】按钮，在弹出的对话框中选择要部署的项目名称和容器名称，单击【OK】按钮，完成项目部署，如图 1-20 所示。

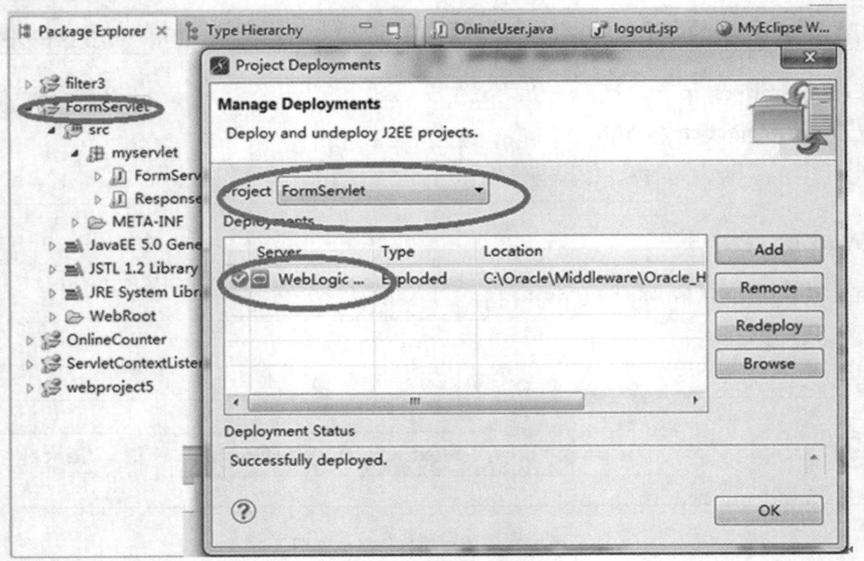

图 1-20　部署项目示意图

项目部署之后，可以在浏览器直接输入：http://服务器地址：服务端口/项目名/网页文件名，例如，http://loaclhost:7001/Helloworld/index.jsp，即可对页面进行访问。如图 1-21 所示。

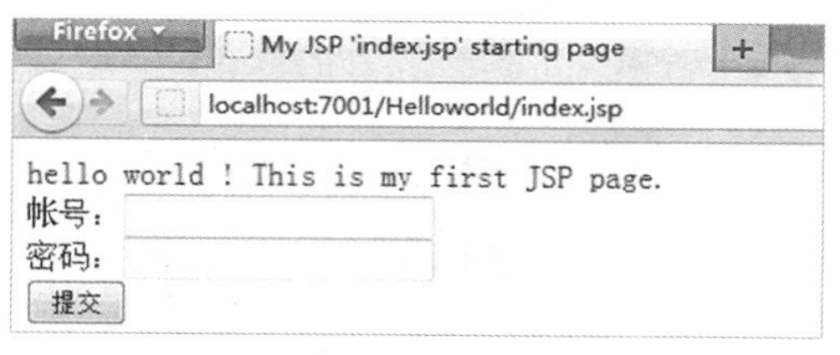

图 1-21　访问页面示意图

1.5　实验总结

本实验的内容是 Java Web 初学者应该掌握的，包括：Java Web 开发环境的搭建和测试，Java Web 项目的创建，Java 程序的编辑与编译，Java Web 程序的设计、部署和运行。要搭建 Java Web 开发环境，首先要下载 JDK、MyEclipse 和容器（WebLogic），然后分别进行安装和配置。安装 JDK 之后，需要配置有关环境变量，例如：JAVA_HOME、PATH、CLASSPATH 等。安装 WebLogic 之后，需要在 WebLogic 控制台进行基础域（Base_Domain）、用户名和密码的设置，并且将 WebLogic 整合到 MyEclipse 中，还需要为 WebLogic 绑定 JDK 环境。新建一个 Java Web 项目，对开发环境进行测试。新建 Web 项目、填写项目信息、JSP 编写页面、部署项目、启动服务器、在浏览器直接输入：http://服务器地址:服务端口/项目名/网页文件名。通过上述步骤搭建并配置好开发环境之后，为将来的 Java Web 程序开发学习打下良好基础，从此可以循序渐进地开展 Java Web 开发的学习。

1.6　课后思考题

1.Web 应用程序使用三层体系结构，具体分为哪三层？

2.Web 请求的方法有哪些？

3.WebLogic 默认服务端口是什么，Tomcat 默认服务端口是什么？

4.描述 C/S 和 B/S 架构的区别，至少写出五个方面。

5.J2EE 是技术、平台还是框架？

6.简述在 MyEclipse 平台下进行 Web 应用程序开发的过程。

实验 2 Java RMI 远程方法调用

2.1 实验目的

1.掌握远程方法调用
2.熟悉 RMI 架构及开发步骤
3.理解 Stub 和 Skeleton 的作用

2.2 实验环境

1.JDK(Java 开发工具包)
2.RMI 通信接口

2.3 实验知识背景

2.3.1 本地计算与分布式计算

本地计算:组件共享一个公共的地址空间进行通信。

分布式计算:组件调用不同机器上的地址空间的组件进行通信。

在实际开发中,如果项目比较庞大,所需要的计算量非常大时,单采用本地计算的方式显然是不够的,不但效率非常低,所需要的时间也很多。那么能否使运行在一台计算机上的程序能够调用位于另一台计算机上的对象呢?

这种情况可以使用分布式计算。分布式计算,简单来说就是把一个需要大量计算的工作分割成多个小块,然后由多台计算机各自运算,再将结果统一合并得出数据结论。

这类的编程通常称为客户机/服务器编程。客户端程序进行方法调用,服务器提供远程对象。当前用于开发分布式对象的一些重要基础结构有：

(1)Java 远程方法调用(RMI)

(2)公共对象请求代理体系结构(CORBA)

(3)Microsoft 分布式组件对象模型(DCOM)

(4)Enterprise JavaBeans(EJB)技术

在本教程中，我们将重点讨论 RMI 技术。

2.3.2 分布式计算角色与机制

在分布式计算中最主要的两个角色是客户端和服务器，如图 2-1 所示。

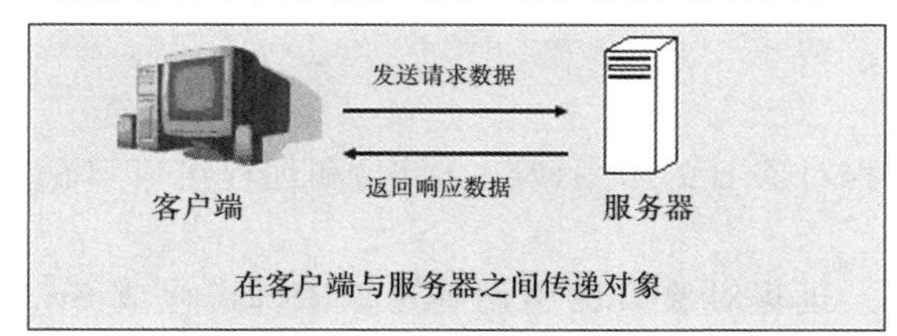

图 2-1 分布式计算关系图

分布式计算机制如下：

(1)客户端以常规的方式进行方法调用。需要考虑提供服务的对象是否在同一个虚拟机内，以及是否为 Java 语言实现的对象。

(2)在客户端为服务器对象安装一个代理(Proxy)，客户端调用此代理进行常规方法调用。客户端代理负责与服务器进行联系。

(3)在服务器端安装第二个代理对象。服务器端代理与客户端代理进行通信，以常规方式调用服务器上的对象的方法。

2.3.3 远程方法调用

当今主要的通信技术包括以下三种：

(1)RMI：Java 的远程方法调用技术，支持 Java 的分布式对象之间的方法调用。

(2)CORBA：通用对象请求代理架构，支持任何编程语言编写的对象之间的方法调用，使用 Internet Inter-ORB Protocol(IIOP，互联网内部对象请求代理协议)支持对象间通信。

(3)SOAP：简单对象访问协议，使用基于 XML 的传输格式。

接下来将详细介绍 Java RMI 架构是如何实现远程方法调用的，如图 2-2 所示。

下面介绍每一个模块的具体作用。

1.服务器端

远程接口 B：列出了可以远程调用的所有方法。

远程对象 B：实现远程接口 B 的类实例化对象。

主干(Skeleton)：将客户端发送的参数反向序列化并调用远程对象上所需的方法，将获得的返回值序列化发送回服务器通信模块。

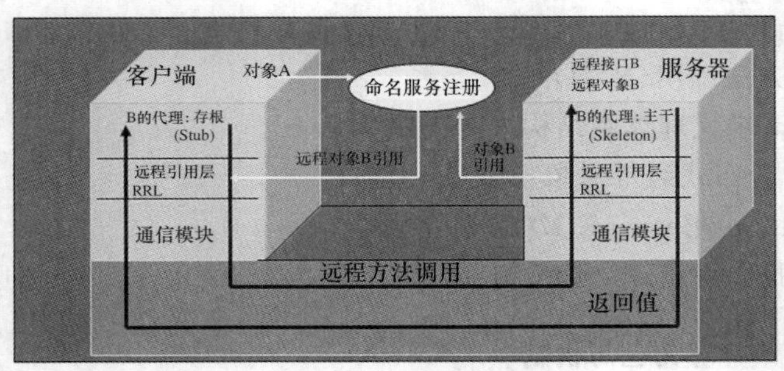

图 2-2　Java RMI 架构

远程引用层(RRL)：为远程对象 B 创建一个远程对象引用，同时维护该远程对象与其引用之间的映射。

2.客户端

存根(Stub)：以远程对象 B 的引用(唯一标识)和远程接口 B 的方法描述构造一个代理对象。

远程引用层(RRL)：远程对象 B 的引用到达 RRL 时，生成 Stub，并维护远程对象 B 的引用与 Stub 之间的映射。

简单地说，RMI 服务器包含了远程调用的方法的对象，并且服务器不仅创建了远程对象，还要在 RMI 注册表中产生这些对象的引用。而 RMI 客户端就可以通过寻找这个对象名在 RMI 注册表中得到一个或多个远程对象的引用，然后通过调用这些远程对象上的方法来访问远程对象的服务。

2.3.4　配置远程方法调用

通过以上讲述，对 RMI 有了初步了解，下面介绍开发 RMI 的步骤。

1.服务器端

(1)定义远程接口，继承 Remote 接口(java.rmi 包)，声明远程调用的方法，每一方法必须声明抛出 RemoteException 异常。

(2)编写实现远程接口的类(服务器类)，该类的对象就是远程对象。

①实现远程接口。

②继承服务器类：java.rmi.server.UnicastRemoteObject。

(3)使用 rmic 工具，将上述服务器类生成存根和主干。

(4)编写一个主类，实例化服务器类，生成远程对象。

(5)命名注册远程对象，利用 java.rmi.Naming 类的方法：

①`public static void bind(String name,Remote obj);`

②`public static void rebind(String name,Remote obj);`

2.客户端

(1)编写客户机类，调用远程对象上的方法。利用 java.rmi.Naming 类的方法：

```
public static Remote lookup(String name);
```

注意：name 以 URL 格式给出：

`rmi://<host_name>:<port>/<service_name>`　(port 默认值为 1099)

(2)启动注册库:start rmiregistry [port]。

(3)运行服务端主类,创建远程对象并向注册库注册该对象引用。

(4)运行客户机。

下面通过一个 circleServer 的例子来更好地体会 RMI 的完整开发过程。如图 2-3 所示。

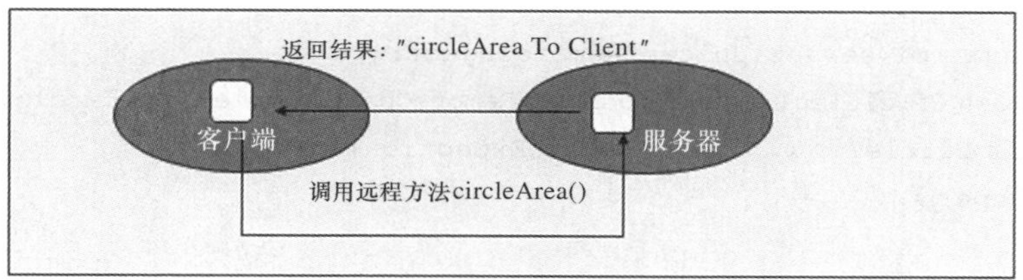

图 2-3 RMI 示例

2.4 实验内容与步骤

1.题目

根据要求编写程序并验证结果,然后回顾本实验内容对结果加以分析总结。

实现一个 RMI 服务器应用程序,提供 circleArea 的服务,根据参数半径获得圆面积:

```
public double circleArea(double r)throws RemoteException;
```

使用一个客户机测试 RMI 服务器,半径参数由客户端从键盘输入。

2.建立项目

打开 MS-DOS 窗口,在 D 盘下创建目录,例如 D:\rmi\homework\src,将.java 源文件保存到该目录下;使用 JDK 工具中 javac 命令编译源文件,使用 rmic-v1.2 服务器类和 start rmiregistry 进行注册及启动 RMI 通信,使用 java 命令运行服务器端和服务器端 java 类文件。项目结构如图 2-4 所示。

```
D:\rmi\homework\src 的目录

2020/12/15  15:47    <DIR>          .
2020/12/15  15:47    <DIR>          ..
2019/03/20  10:55               378 CircleImpl.class
2019/03/18  17:09               344 CircleImpl.java
2019/03/20  10:59             1,772 CircleImpl_Stub.class
2019/03/20  10:55               215 CircleInterface.class
2019/03/18  17:08               208 CircleInterface.java
2019/03/20  10:56               703 CircleServer.class
2019/03/18  19:11               344 CircleServer.java
2019/03/20  10:56             1,266 testCircle.class
2019/03/18  19:25               616 testCircle.java
               9 个文件          5,846 字节
               2 个目录 306,089,226,240 可用字节
```

图 2-4 项目结构示意图

3.定义远程接口

```
package testrmi;
import java.rmi.Remote;
import java.rmi.RemoteException;
```

```
public interface CircleInterface extends Remote{
    public double circleArea(double r) throws RemoteException;
}
```

4.定义服务器类

```
package testrmi;
import java.rmi.RemoteException;
import java.rmi.server.UnicastRemoteObject;
public class CircleImpl extends UnicastRemoteObject implements CircleInterface{
    public CircleImpl() throws RemoteException{
        super();
    }
    public double circleArea(double r)throws RemoteException{
        return 3.14*r*r;
    }
}
```

5.创建 RMI 服务器主类,远程对象注册

```
package testrmi;
import java.rmi.Naming;
public class CircleServer{
    public static void main(String args[]){
        try{
            CircleImpl circle=new CircleImpl();
            Naming.rebind("rmi://localhost:1099/circle_area",circle);
            System.out.println("object bound");
        }catch(Exception e){
            e.printStackTrace();
        }
    }
}
```

6.创建客户端程序

```
package testrmi;
import java.io.BufferedReader;
import java.io.InputStreamReader;
import java.rmi.Naming;
public class testCircle{
    public static void main(String args[]){
        try{
            CircleInterface cir=(CircleInterface)
                    Naming.lookup("rmi://localhost:1099/circle_area");
            System.out.println("请输入半径:");
            BufferedReader br=new BufferedReader(new InputStreamReader(System.in));
            double r=Double.parseDouble(br.readLine());
            System.out.println("Circle Area:"+cir.circleArea(r));
```

```
        }catch(Exception e){
            e.printStackTrace();
        }
    }
}
```

7.运行服务器端程序并注册

运行服务器端程序并注册,如图 2-5 所示。

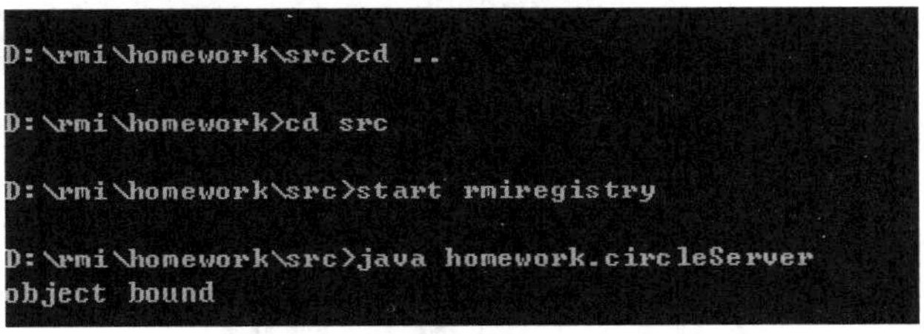

```
D:\rmi\homework\src>cd ..

D:\rmi\homework>cd src

D:\rmi\homework\src>start rmiregistry

D:\rmi\homework\src>java homework.circleServer
object bound
```

图 2-5　运行服务器端程序并注册

8.运行客户端程序并输入半径值

运行客户端程序并输入半径值,如图 2-6 所示。

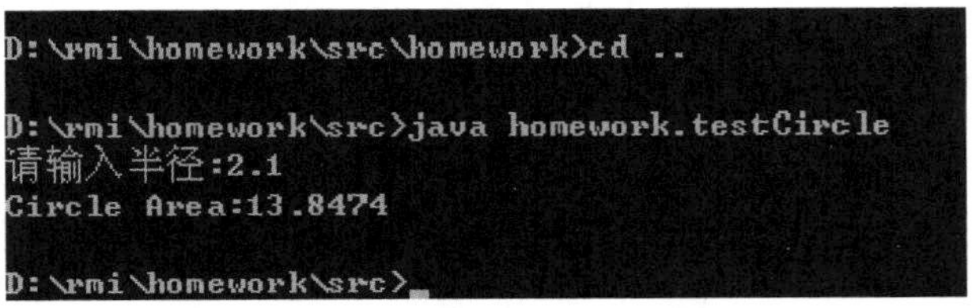

```
D:\rmi\homework\src\homework>cd ..

D:\rmi\homework\src>java homework.testCircle
请输入半径:2.1
Circle Area:13.8474

D:\rmi\homework\src>
```

图 2-6　运行客户端程序

2.5 实验总结

通过本次实验理解了本地计算与分布式计算的基本概念和计算方法,了解了分布式计算的角色与机制,掌握远程方法调用(Remote Method Invocation,RMI)以及如何配置远程方法调用,了解远程方法中的参数传递,以及 RMI 与 CORBA 区别,初步建立 Web Application 及企业计算概念,掌握分布式计算应用程序的设计步骤和方法。

2.6 课后思考题

1.java.rmi.Naming 类采用何种方法向注册库注册远程对象?

2.RMI 注册类库实现何种接口?

3.简单描述本地计算和分布式计算的区别。

4.简述分布式计算的三种通信技术。

5.简述使用 RMI 创建 C/S 应用程序的步骤。

实验 3

JNDI 与数据源

3.1 实验目的

1.了解 JNDI 命名服务基础
2.掌握在 WebLogic Server 环境下使用 JNDI 命名服务的方法
3.了解描述数据源及 JDBC 中的主要接口
4.了解 WebLogic Server 环境对 JDBC 的支持

3.2 实验环境

1.MyEclipse 插件平台
2.WebLogic(或 Tomcat)容器
3.MySQL 数据库

3.3 实验知识背景

3.3.1 命名和目录服务

命名服务:提供一种绑定服务,即映射名称到某个对象以及为客户端提供接口,通过名称访问对象的服务。例如:RMI、DNS 等。

目录服务:特殊的命名服务,在建立名字到对象的映射时还可以设置更多的属性。目录服务安排在层次化树状结构中。例如:NDS、LDAP 等。

名称与对象的关联处理称为绑定。一组绑定又可以称为上下文,有助于对绑定进行分类,简化对象的搜索。上下文内部可以有另一个上下文,嵌入的上下文就是子上下文。

子上下文仅在目录服务中有效，而在命名服务中是无效的，这是因为命名服务未安排在层次化树状结构中。

注意：目录服务是一个命名服务，命名服务不一定是目录服务。

3.3.2 JNDI 架构和 API

Java 命名与目录接口(Java Naming and Directory Interface,JNDI)为 Java 程序提供访问命名与目录服务的 API。可以使用 JNDI 获得对现有命名服务和新创建的命名服务的访问，还可以通过 JNDI 的服务提供程序接口(SPI)集成更多不同种类的命名服务。

JNDI 架构提供了一组标准的独立于命名系统的 API，这些 API 构建在与命名系统有关的驱动之上。这一层有助于将应用层与实际数据源分离，无论应用访问的是 LDAP、RMI、DNS 还是其他的目录服务。总之，JNDI 是一个独立于目录服务的具体实现，只要有目录的服务提供接口(或驱动)即可使用目录。如图 3-1 所示。

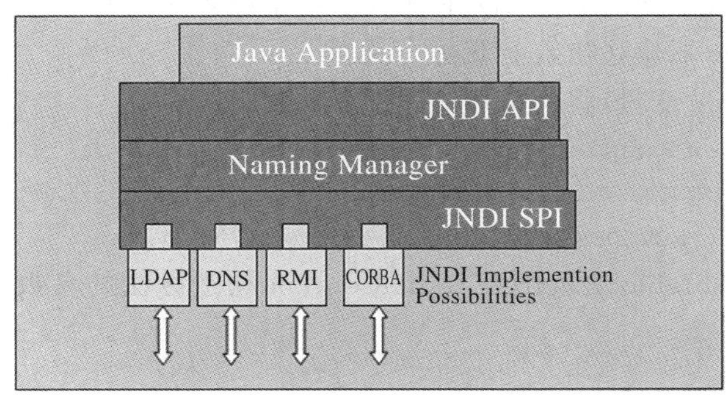

图 3-1 JNDI 架构

JNDI 包含 5 个包及一个接口 Context，如表 3-1 所示。

表 3-1 JNDI 主要包及接口

javax.naming	访问命名服务的类与接口
javax.naming.event	在命名服务中实现事件通知机制的类与接口
javax.naming.ldap	支持 LDAP v3 扩展和控制的类与接口
javax.naming.spi	JNDI 的服务提供程序接口(SPI)，一般用户不会涉及
javax.naming.directory	访问目录服务的类与接口
Context	接口，定义命名服务的基本操作及创建子上下文等操作

JNDI Context 接口定义许多命名服务操作：

```
public interface Context{
    public Object lookup(String name) throws NamingException;
    public void bind(String name,Object obj) throws NamingException;
    public void rebind(String name,Object obj) throws NamingException;
    public void unbind(String name) throws NamingException;
    public void rename(String oldname,String newname) throws NamingException;
    public Context createSubcontext(String name) throws NamingException;
```

```
    public void destroySubcontext(String name) throws NamingException;
    ......
}
```

javax.naming.InitialContext 类实现 Context 接口。

3.3.3 在 WebLogic Server 环境下使用 JNDI 的命名服务

接下来将详细讲解如何在 WebLogic Server 环境下使用 JNDI。首先来介绍如何创建初始上下文。

InitialContext 类实现 Context 接口，调用其构造方法便可创建初始上下文(即搜索请求对象的起始点)。创建初始上下文需要设定以下两个环境属性：

①上下文工厂对象 spi Context.INITIAL_CONTEXT_FACTORY，指定要使用哪个具体服务提供程序。

②连接字符串 Context.PROVIDER_URL，指定服务的位置和初始上下文的起始点。

(1)在 WebLogic 下建立初始上下文

①建立 Hashtable 变量，将两个环境属性存入其中。

```
Hashtable ht=new Hashtable();
ht.put(Context.INITIAL_CONTEXT_FACTORY,"weblogic.jndi.WLInitialContextFactory");
ht.put(Context.PROVIDER_URL,"t3://localhost:7001");
```

②使用上述 Hashtable 变量作为 InitialContext 构造方法的参数，创建一个 Context 实例。

```
try{
    Context ctx=new InitialContext(ht);
}
```

(2)绑定和查找对象

①绑定对象：调用 Context 接口中的 bind()和 rebind()方法来实现。

```
Context.bind(String name,Object obj);
Context.rebind(String name,Object obj);
```

②查找对象：调用 Context 接口中的 lookup(String name)方法返回当前上下文中参数 name 对应的绑定对象。

```
Object Context.lookup(String name);
```

3.3.4 描述数据源及 JDBC 中的主要接口

数据源(Data Source)：数据的来源，提供所需要数据的器件或原始媒体。如同通过指定文件名，便可以在文件系统中找到文件一样，通过提供正确的数据源名称，就可以找到相应的数据库连接，获取数据。

JDBC(Java 数据库连接)是 Java 程序与数据库系统通信的标准 API，由一系列连接(Connection)、SQL 语句(Statement)和结果集(ResultSet)构成。如图 3-2 所示。

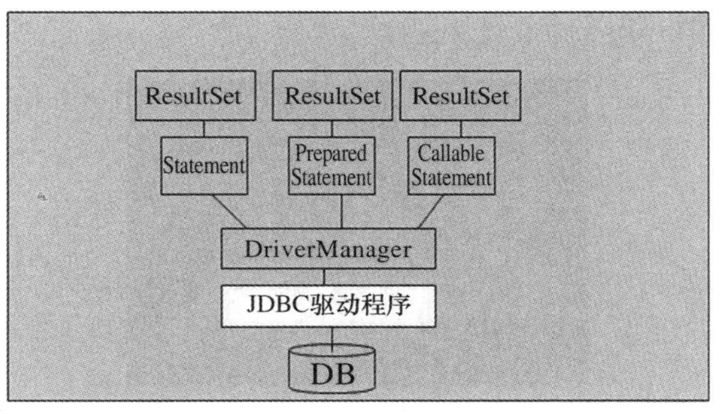

图 3-2　JDBC 结构

JDBC 中的主要接口如表 3-2 所示。

表 3-2　　JDBC 主要接口

接　口	作　用
java.sql.DriverManager	处理驱动程序的加载和建立新数据库连接
java.sql.Connection	处理与特定数据库的连接
java.sql.Statement	在指定连接中处理 SQL 语句
java.sql.ResultSet	处理数据库操作结果集

3.3.5　WebLogic Server 环境对 JDBC 的支持

WebLogic Server 数据源配置步骤如下：

(1)建立数据源(DataSource)。数据源指明连接数据库的一些信息，如数据库的 JDBC 驱动程序、数据库名称、帐号与密码、数据源对象(可以通过 JNDI 查询到)。在 WebLogic Server 管理控制台下完成。

(2)访问数据库时通过数据源的 JNDI 名称查找该数据源，然后通过数据源获得 Connection 对象。

下面通过代码进行详细说明。

(1)建立数据库表 user，如图 3-3 所示。

[表设计] user @test (loc...
文件(F)　编辑(E)　窗口(W)
创建　保存　另存为　创建栏位　插入栏位　删除栏位　主键　上移　下移
栏位　索引　外键　触发器　选项　注记　SQL 预览

名	类型	长度	十进位	允许空?..	
userid	int	11	0	☐	1
username	varchar	50	0	☑	

图 3-3　建立数据库表

（2）查询数据库数据，如图 3-4 所示。

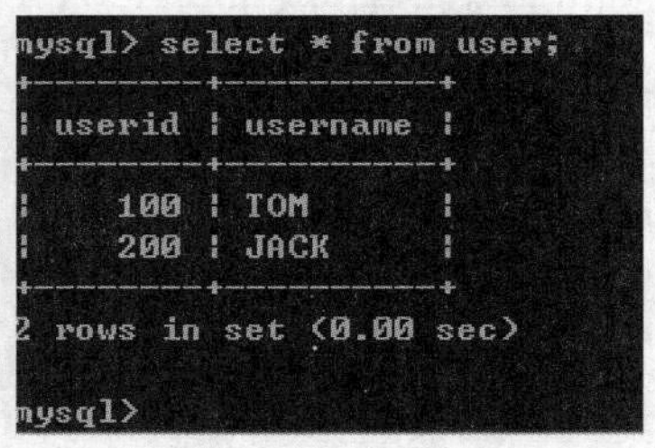

图 3-4　查询数据库数据

（3）建立 JDBC 数据源。

开启 WebLogic，用浏览器打开 http://localhost:7001/console，单击【JDBC】|【DATA Source】|【New】，创建数据源，填写相关信息（以下截图仅截取主要步骤），如图 3-5 所示。

图 3-5　建立 JDBC 数据源

（4）填写数据库连接信息，如图 3-6 所示。

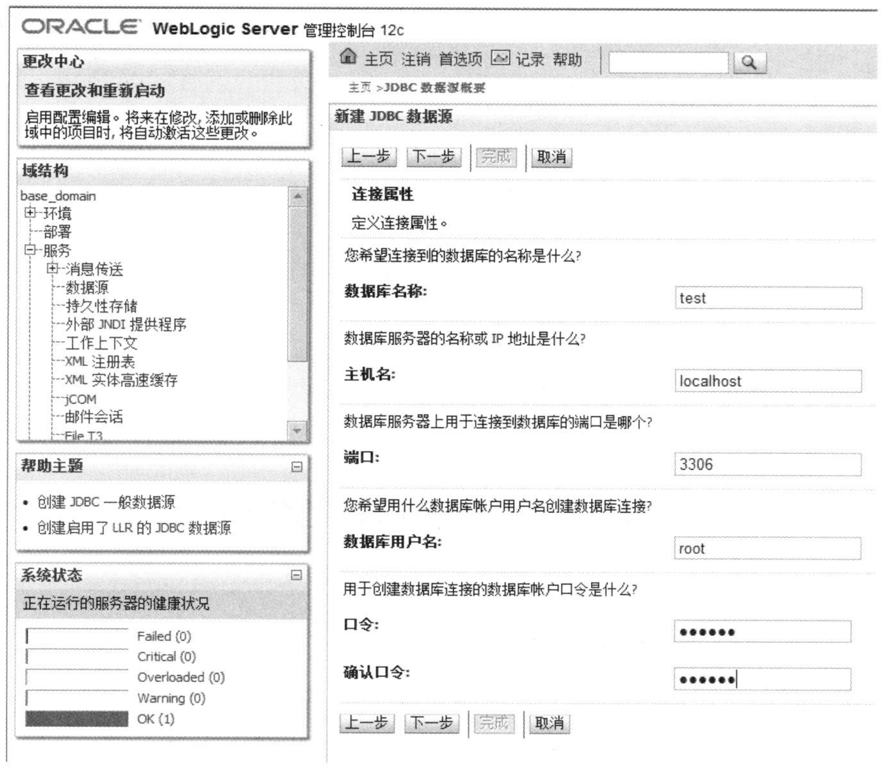

图 3-6　填写数据库连接信息

(5)测试连接成功后系统将给出提示,如图 3-7 所示。

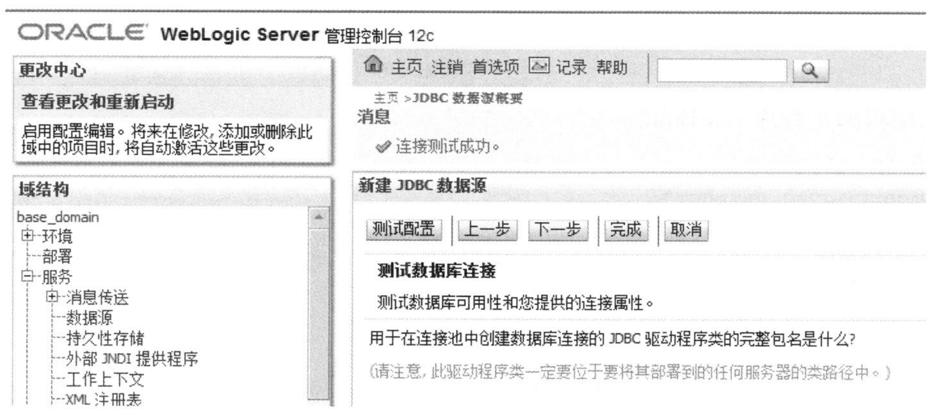

图 3-7　测试连接成功提示

在完成设置之前,用"√"符号选择 AdminServer 选项完成 JDBC 数据源的设置。如图 3-8 所示。

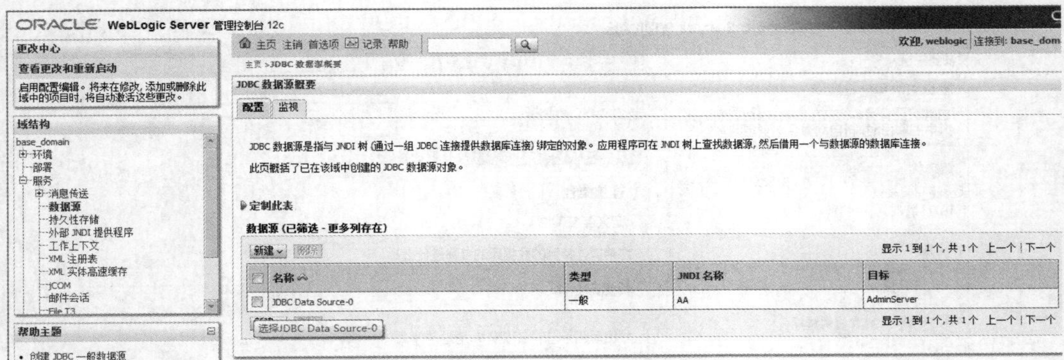

图 3-8　选择 AdminServer 选项

3.4　实验内容与步骤

3.4.1　对象绑定及 JNDI 查找操作

1.实验原理如图 3-9 所示。

图 3-9　实验原理图

2.编写绑定程序 testBind.java。

```
package testjndi;
import javax.naming.*;
import java.util.*;
//将对象绑定到 WebLogic Server 的命名服务中
public class TestBind{
    public static void main(String[] args){
        Hashtable ht=new Hashtable();
        ht.put(Context.INITIAL_CONTEXT_FACTORY,"weblogic.jndi.WLInitialContextFactory");
        ht.put(Context.PROVIDER_URL,"t3://localhost:7001");
        try{
            Context ctx=new InitialContext(ht);
            String str="Hello world !";
            ctx.bind("hello",str);
```

```
            System.out.println("Object Bound !");
        }catch(Exception e){
            e.printStackTrace();
        }
    }
}
```

3.编写查找对象程序 testLookup.java。

```
package testjndi;
import javax.naming.*;
import java.util.*;
//通过 JNDI 查找指定的对象
public class TestLookup{
    public static void main(String[] args){
        System.out.println("Object Binding...");
        Hashtable ht=new Hashtable();
        ht.put(Context.INITIAL_CONTEXT_FACTORY,"weblogic.jndi.WLInitialContextFactory");
        ht.put(Context.PROVIDER_URL,"t3://localhost:7001");
        try{
            Context ctx=new InitialContext(ht);
            String result;
            result=(String)ctx.lookup("hello");
            System.out.println("Result is "+ result);
            System.out.println("Lookup end !");
        }catch(Exception e){
            e.printStackTrace();
        }
    }
}
```

4.为了方便更改环境属性,把环境属性添加到属性文件中。

```
java.naming.factory.initial=weblogic.jndi.WLInitialContextFactory
java.naming.provider.url=t3\://localhost\:7001
```

5.编写绑定对象程序 TextBind_Property.java。

```
package testjndi;
import javax.naming.*;
import java.io.*;
import java.util.*;
public class TextBind_Property{
    public static void main(String[] args){
        try{
            Properties props=new Properties();
            props.load(new FileInputStream(".\\bin\\testjndi\\TextBind_Property.
```

```
        properties"));
        Context ctx=new InitialContext(props);
        String str="Hello Enterprise java !";
        ctx.bind("hello",str);
        System.out.println("Object Bound !");
    }catch(Exception e){
        e.printStackTrace();
      }
  }
}
```

6.在构建路径下为项目添加 weblogic.jar 包支持,如图 3-10 所示。

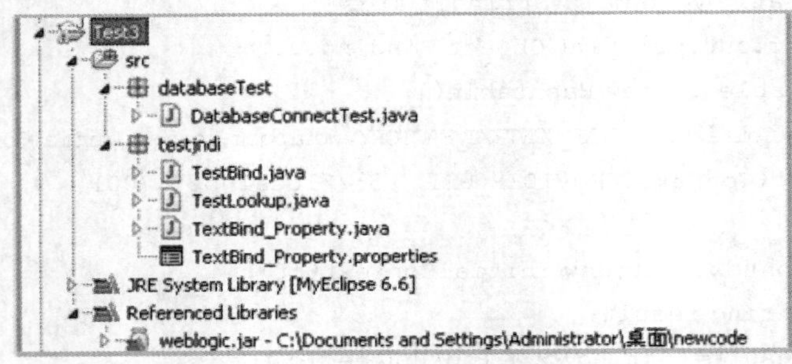

图 3-10 项目添加 weblogic.jar 包

7.运行 TextBind_Property.java 程序,程序运行结果如图 3-11 所示。

图 3-11 TextBind_Property.java 程序运行结果

8.运行 TestLookup.java 程序,程序运行结果如图 3-12 所示。

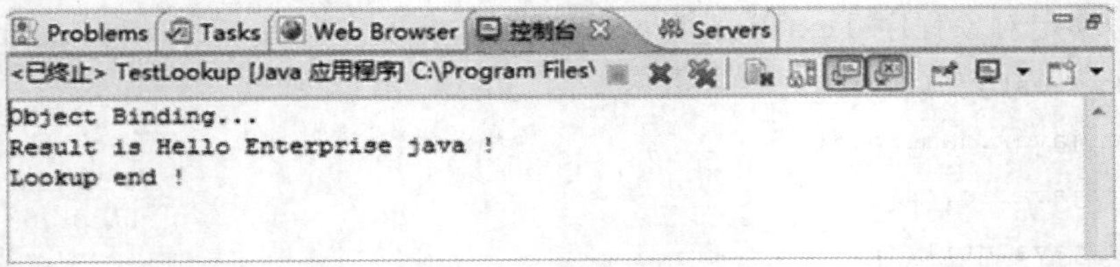

图 3-12 TestLookup.java 程序运行结果

3.4.2 JNDI 查找数据源

1.实验原理如图 3-13 所示。

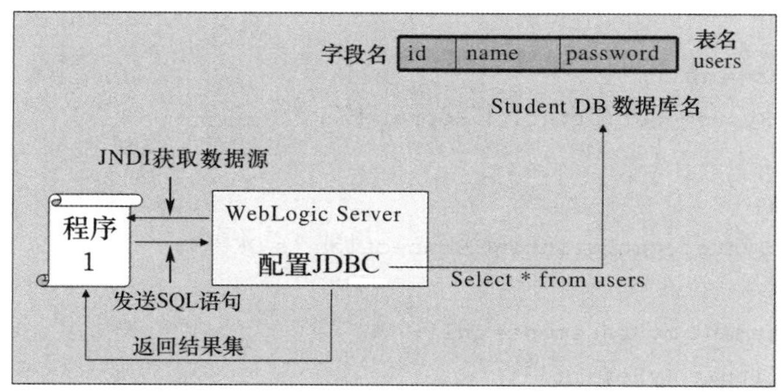

图 3-13 实验原理图

2.添加 WebLogic Server 环境对 JDBC 的支持，在 MySQL 上建立 student 数据库，创建 student 表。如图 3-14 所示。

图 3-14 student 数据库

3.在 WebLogic 中建立 JDBC 数据源，例如 JavaEE_HomeWork2。

4.编写数据库连接测试类 DatabaseConnectTest.java。

```
package databaseTest;
import javax.naming.*;
import javax.sql.*;
import java.sql.*;
import java.util.*;
public class DatabaseConnectTest{
    public static void main(String args[]){
        DataSource ds=null;
        Context ctx;
        Connection myConn=null ;
        Hashtable ht=new Hashtable();
        ht.put(Context.INITIAL_CONTEXT_FACTORY,"weblogic.jndi.WLInitialContextFactory");
        ht.put(Context.PROVIDER_URL,"t3://localhost:7001");
        try{
```

```
            ctx=new InitialContext(ht);
            ds=(javax.sql.DataSource)ctx.lookup("JavaEE_HomeWork2");
        }catch(Exception e){
            e.printStackTrace();
        }
        if(ds==null){
            System.out.println("Eorror !");
        }
        else{
            System.out.println("Connection is OK !");
        }
        Statement myStatement=null;
        ResultSet mySet=null;
        try{
            myConn=ds.getConnection();
            myStatement=myConn.createStatement();
            mySet=myStatement.executeQuery("select* from student;");
            while(mySet.next()){
                System.out.println(mySet.getString("studentID")+ "\t"+
                mySet.getString("studentName"));
            }
        }catch(Exception e){
            e.printStackTrace();
        }
    }
}
```

5.运行 DatabaseConnectTest.java，测试结果如图 3-15 所示。

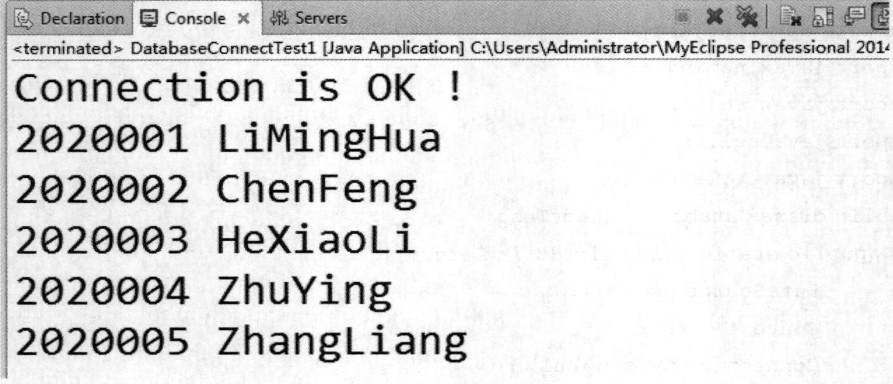

图 3-15　程序运行结果

3.5 实验总结

通过本次实验的学习，充分理解命名和目录服务的定义、区别，并且熟练掌握 JNDI 架构及其 API。通过在 WebLogic Server 环境下使用 JNDI 命名服务的例子，编写查询数据的相应程序，更好地掌握 WebLogic Server 环境对 JDBC 的支持。特别要注意，目录服务是一个命名服务，命名服务不一定是目录服务。

3.6 课后思考题

1.什么是命名或目录服务中的所有名称的集合？

2.当从连接池获得某个连接对象，并在该对象上调用 close()方法时，连接是如何实现的？

3.简要描述 Web Service 的涵义。

4.如何建立初始上下文？

5.简要描述数据源及 JDBC 中的主接要口。

6.简要描述 WebLogic Server 下的 JDBC 开发步骤。

实验4 JavaBean构件设计

4.1 实验目的

1.掌握 JavaBean 的设计、命名规范
2.掌握 Bean 的定制事件方法
3.掌握 Bean 属性的绑定方法
4.掌握 Bean 属性的约束属性的设计方法
*5.了解 BeanInfo 类的设计过程和方法

4.2 实验环境

1.JDK(Java 开发工具包)
2.MyEclipse(插件平台)

4.3 实验知识背景

1.JavaBean 编写规范

本质上 JavaBean 是 Java 类,但是为了实现 JavaBean 的特定功能,编写 JavaBean 除了要遵守一般 Java 类的编写要求之外,还有一些特定的要求,可以称其为 JavaBean 编写规范,或 JavaBean 规范。Sun 公司提供的 JavaBean 规范可在 Oracle 公司网站下载,网址是:http://www.oracle.com/technetwork/java/javase/documentation/spec-136004.html。

Sun 公司的 JavaBean 规范内容繁多,在实际编程中可以逐步学习、理解和应用。下面是 JavaBean 编程中要遵守的几个常用规范:

(1)JavaBean 必须包含一个无参数的 public 构造方法;

(2)JavaBean 必须包含符合命名规范的 get 和 set 方法;

(3)JavaBean 应该是可序列化的,实现 Serializable 接口;

(4)JavaBean 必须支持内省;

(5)如果 Bean 有运行时外观,就必须扩展 java.awt.ComPonent 类。

2.JavaBean 属性、方法与事件

JavaBean 属性通常应该遵循简单的方法命名规则,这样应用程序构造器工具和最终用户才能找到 JavaBean 提供的属性,然后查询或修改属性值,对 Bean 进行操作。

JavaBean 中的方法就是普通 Java 类的方法,可以被其他组件或脚本调用。默认情况下,所有 Bean 的公有方法都可以被外部调用。JavaBean 严格遵守面向对象的类设计逻辑,不让外部对象访问其任何字段(没有 public 字段)。方法调用是接触 Bean 的唯一途径。

事件为 JavaBean 组件提供了一种发送通知给其他组件的方法。在 AWT 事件模型中,一个事件源可以注册事件监听器对象。当事件源检测到发生了某种事件时,将调用事件监听器对象中相应的事件处理方法来处理这个事件。

JavaBean 的属性与一般 Java 程序中所指的类属性,或者说与所有面向对象的程序设计语言中对象的类属性是相同的概念,在程序中的具体体现就是类中的变量。在 JavaBean 的设计中,按照属性的作用不同又细分为四类:单值属性、索引属性、关联属性和限制属性。

(1)JavaBean 中的单值属性

①含义

单值属性表示一个伴随有一对 get/set 方法的变量。属性名与和该属性相关的 get/set 方法名对应。

②编码规则

- 属性声明为私有,没有公有的属性;
- 通过方法访问而不是直接存储实例变量;
- 属性值 xxx 对应有 getXxx 和 setXxx 方法;
- 对于布尔型的属性,可以用 isXxx 方法查值。

Bean 的普通方法不用遵循上面的命名规则,但这些方法是公有的。

(2)JavaBean 中的索引属性

JavaBean 中的索引属性表示一个数组值。使用与该属性对应的 get/set 方法可取得数组中的成员数据值(需要有一个整数索引参数),一次设置或取得整个数组的值。

(3)JavaBean 中的关联属性

JavaBean 中的关联属性是指当该种属性的值发生变化时,要通知其他对象。每次属性值改变时,将触发一个 PropertyChange 事件。事件中封装了属性名、属性的原值和属性变化后的新值。至于接收事件的 Bean 应做什么动作由其自己定义。包含关联属性的 Bean 必须具有以下功能:

①允许事件监听器注册和注销与其有关的属性修改事件;

②当修改一个关联属性时,可以在相关的监听器上触发属性修改事件。

在 java.beans 包中利用 PropertyChangeSupport 类创建该类的对象，从而可以用于管理注册的监听器列表和属性修改事件通知的发送。

JavaBean 中关联属性的编程要点如下：

①当关联属性被改变时，将使用 PropertyChangeSupport 类中的 firePropertyChange 方法来触发属性改变事件。

②需要存储关联属性的原始属性值，因为原始属性值和新的属性值都要传给 firePropertyChange 方法，而且均为 Object 类型。如果为基本类型的数据，则应该转换为对应的类型。

③属性修改事件是在属性被修改后触发。

```
changes.firePropertyChange("ourString",oldString,newString);
```

④Bean 要预留出一些接口给开发工具，开发工具使用这些接口，把其他的 JavaBean 对象与该 Bean 相连接。

(4)JavaBean 中的限制属性

JavaBean 中的限制属性指当该属性的值要发生变化时，与这个属性已建立了某种连接的其他外部 Java 对象可否决该属性值的改变，Bean 本身也可以否决该属性值的改变。

限制属性的监听者通过抛出 PropertyVetoException 异常来阻止该属性值的改变。

限制属性一般有两种监听者：属性变化监听者和否决属性改变的监听者。否决属性改变的监听者在自己的对象代码中有相应的控制语句，在监听到有限制属性要发生变化时，在控制语句中判断是否应否决该属性值的改变。

限制属性的编程要点：

①分别声明 PropertyChangeSupport 和 VetoableChangeSupport 类的对象(限制属性值可否改变)。

②当属性被改变后触发属性改变否决事件。

③若有其他对象否决属性的改变，程序将抛出异常，不再继续执行后续的语句，方法结束；若无其他对象否决属性发生改变，则把该属性赋予新值，并触发属性改变事件。

④与属性改变事件相同，也要为限制属性预留接口，使其他对象可注册到限制属性的否决改变监听者队列中，或把该对象从中注销。

⑤限制属性实际上是一种特殊的关联属性，只是其值的变化可以被监听者否决。

3.Java 内省和 BeanInfo 接口

Java 内省(Java Introspector)是 Java 语言对 JavaBean 属性、事件的一种处理方法，也就是说给定一个 JavaBean 对象，就可以得到或调用它的所有 get/set 方法。例如，类 A 中有属性 name，可以通过 getName、setName 方法来得到其值或者设置新值。通过 getName、setName 方法来访问 name 属性，这是默认的规则。Java 内省机制可以使用另一种方式访问 JavaBean 某个属性的 getter、setter 方法，并获得 JavaBean 的方法名称。

一般的做法是通过类 Introspector 来获取某个对象的 BeanInfo 接口信息，然后通过 BeanInfo 接口来获取属性的描述器(Property Descriptor)，通过这个属性描述器可以获取某个属性对应的 getter、setter 方法，然后就可以通过反射机制来调用这些方法。

4.4 实验步骤与内容

4.4.1 实验步骤

1.创建一个 Java 项目。单击【文件】|【新建】|【Java 项目】，在弹出的窗口中填写项目名称等信息，单击【确定】按钮。

2.创建绑定类文件，将其保存在源目录路径下(缺省为.\src)。展开项目名称文件，选择包目录名(如 src)，右击选择【新建】|【Java 类】，在弹出的窗口中填写类名称等信息，单击【确定】按钮。

3.创建绑定类的属性文件，将其保存在源目录路径下(缺省为.\src)。展开项目名称文件，选择包目录名(如 src)，右击选择【新建】|【File】，在弹出的窗口中填写属性名称，如 xxx.properties，填写完整其他信息，单击【确定】按钮。

4.将 weblogic.jar 构建路径到项目库中。具体操作是右击项目名称，在弹出的菜单中选择“属性”选项，在弹出的界面中打开“库”界面，单击【添加外部 jar】按钮，找到 WebLogic 安装路径(C:\bea\webserv_10.0\server\lib\weblogic.jar)下的 weblogic.jar，单击【打开】|【确定】按钮，将 weblogic.jar 构建路径到项目库中。

5.在 MyEclipse 的 server 选项中，启动 WebLogic 容器，确认其处于“Run”状态。

6.选择 Java 主程序名称右击，在弹出的菜单中选择【Run As】|【Java Application】，在 MyEclipse Console 中观察程序运行结果。

4.4.2 实验内容

1.设计一个学生 Bean，当设置学生的年龄大于 30 岁时，该 Bean 产生一个 Event，输出提示信息：“年龄大于 30 岁，你目前的年龄是：”。

(1)项目结构如图 4-1 所示。

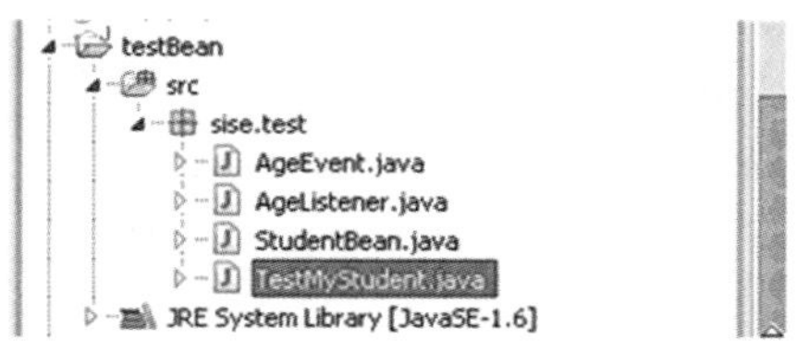

图 4-1 项目结构图

(2)定义事件 AgeEvent，继承类 EventObject。

```
package sise.test;
import java.util.EventObject;
public class AgeEvent extends EventObject{
    private static final long serialVersionUID=1L;
```

```
    String msg;
    public AgeEvent(Object source,String msg){
        super(source);
        setMsg(msg);
    }
    public String getMsg(){
        return msg;
    }
    public void setMsg(String msg){
        this.msg=msg;
    }
}
```

(3)定义监听接口 AgeListener,继承 EventListener 接口,其中 ageCheck 方法用来检查学生年龄。

```
package sise.test;
import java.util.EventListener;
public interface AgeListener extends EventListener{
    public void ageCheck(AgeEvent e);
}
```

(4)设计学生 Bean,其中增加 Vector 属性,保存已注册监听程序。增加监听程序的注册/删除方法。当设置年龄时,大于 30 岁将产生 AgeEvent 事件。

```
package sise.test;
import java.beans.PropertyChangeListener;
import java.beans.PropertyChangeSupport;
import java.beans.PropertyVetoException;
import java.beans.VetoableChangeListener;
import java.beans.VetoableChangeSupport;
import java.io.Serializable;
import java.util.Enumeration;
import java.util.Vector;
public class StudentBean implements Serializable{
    private static final long serialVersionUID=1L;
    private String name;
    private int age;
    private String university;
    private Vector< AgeListener> listeners;
    PropertyChangeSupport pcs;
    VetoableChangeSupport vcs;
    public StudentBean(){
        pcs=new PropertyChangeSupport(this);//绑定属性
```

```
        vcs=new VetoableChangeSupport(this);//限制属性
        setName("Tom");
        setAge(21);
        setUniversity("GZU");
        listeners=new Vector< AgeListener>();
    }
    public String getName(){
        return name;
    }
    public void setName(String name){
        this.name=name;
    }
    public int getAge(){
        return age;
    }
    public void setAge(int age){
        //this.age=age;
        int old;
        old=this.age;
        try{
            vcs.fireVetoableChange("age",old,age);
            this.age=age;
        }catch(PropertyVetoException e){
            System.out.println(e);
          }
        if(age>=30){
            AgeEvent aev=new AgeEvent(this,"年龄大于 30 岁,"+ "你目前的年龄是:"+age);
            Enumeration< AgeListener> list=getListeners().elements();
            while(list.hasMoreElements()){
                AgeListener listen=(AgeListener)list.nextElement();
                listen.ageCheck(aev);
            }
        }
    }
    public String getUniversity(){
        return university;
    }
    public void setUniversity(String u){
        String old=university;
        this.university=u;
```

```
        pcs.firePropertyChange("university",old,u);//绑定属性
    }
    public Vector< AgeListener> getListeners(){
        return listeners;
    }
    public void setListeners(Vector< AgeListener> listeners){
        this.listeners=listeners;
    }
    public void _toString(){
        System.out.println("Personal Information:"+ name+ ","+ age+ ","+
        university);
    }
    public void addAgeListener(AgeListener al){
        if(al!=null)
            getListeners().add(al);
    }
    public void removeAgeListener(AgeListener al){
        if(al!=null)
            getListeners().remove(al);
    }
    public void addPropertyChangeListener(PropertyChangeListener pcl){
        pcs.addPropertyChangeListener(pcl);
    }
    public void removePropertyChangeListener(PropertyChangeListener pcl){
        pcs.removePropertyChangeListener(pcl);
    }
    public void addVetoableChangeListener(VetoableChangeListener vcl){
        vcs.addVetoableChangeListener(vcl);
    }
    public void removeVetoableChangeListener(VetoableChangeListener vcl){
        vcs.removeVetoableChangeListener(vcl);
    }
}
```

(5)设计测试程序。

```
package sise.test;
import java.beans.PropertyChangeEvent;
import java.beans.PropertyChangeListener;
import java.beans.PropertyVetoException;
import java.beans.VetoableChangeListener;
public class TestMyStudent{
```

```
    public TestMyStudent(){
    }
    public static void main(String[] args){
        StudentBean st=new StudentBean();
        System.out.println("初始信息:");
        st._toString();
        st.addAgeListener(new AgeListener(){
            public void ageCheck(AgeEvent e){
                System.out.println(e.getMsg());
            }
        });
        st.addPropertyChangeListener(new PropertyChangeListener(){
            public void propertyChange(PropertyChangeEvent e){
                System.out.println("University Changed:"+e.getNewValue());
            }
        });
        st.addVetoableChangeListener(new VetoableChangeListener(){
            public void vetoableChange(PropertyChangeEvent e)throws PropertyVetoException{
                if(Integer.parseInt((e.getNewValue()).toString())<=20)
                throw(new PropertyVetoException("年龄太小"+e.getNewValue(),e));
            }
        });
        st.setUniversity("HR");
        st.setAge(19);
        st.setAge(39);
        System.out.println("最后信息:");
        st._toString();
    }
}
```

(6)程序运行结果如图 4-2 所示。

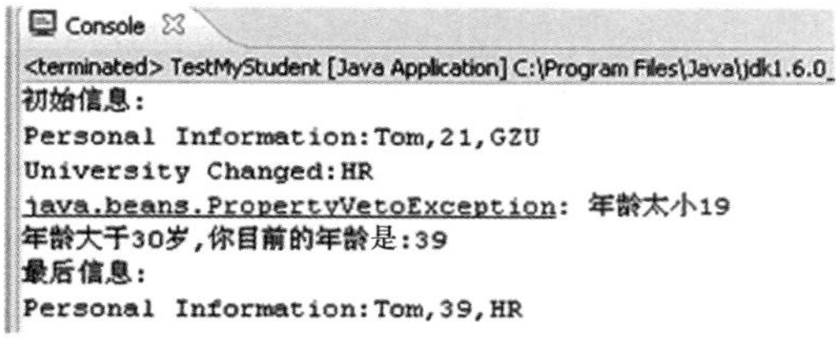

```
Console
<terminated> TestMyStudent [Java Application] C:\Program Files\Java\jdk1.6.0_
初始信息:
Personal Information:Tom,21,GZU
University Changed:HR
java.beans.PropertyVetoException: 年龄太小19
年龄大于30岁,你目前的年龄是:39
最后信息:
Personal Information:Tom,39,HR
```

图 4-2　程序运行结果 1

2.设计学生 Bean,其中 university 属性为绑定属性,当该属性的值发生变化时,输出“university changed”。

(1)绑定属性项目结构如图 4-3 所示。

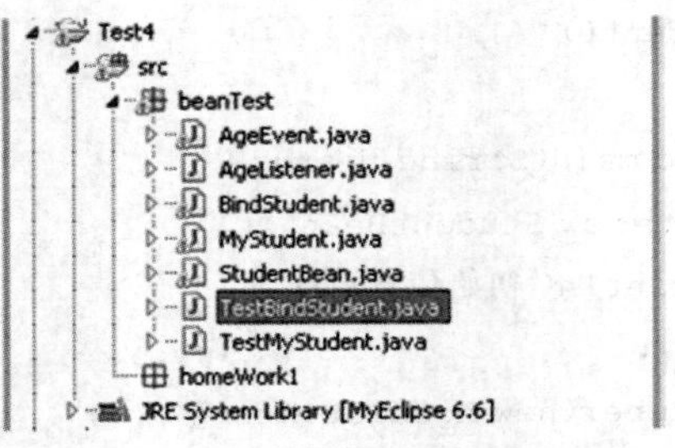

图 4-3　绑定属性项目结构图

(2)定义绑定属性类 BindStudent,实现 Serializable 接口。

```
package beanTest;
import java.io.*;
import java.beans.*;
public class BindStudent implements Serializable{
    String name;
    int age;
    String university;
    PropertyChangeSupport pcs;
    public BindStudent(){
        pcs=new PropertyChangeSupport(this);
        setName("Tom");
        setAge(21);
        setUniversity("GZU");
    }
    public void _toString(){
        System.out.println("Personal Information:"+ name+ "  "+ age+ "  "+
        university);
    }
    public String getName(){
        return name;
    }
    public void setName(String name){
        this.name=name;
    }
    public int getAge(){
        return age;
    }
    public void setAge(int age){
        this.age=age;
    }
    public String getUniversity(){
```

```
        return university;
    }
    public void setUniversity(String sl){
        String old=university;
        university=sl;
        pcs.firePropertyChange("university",old,sl);
    }
    public void addPropertyChangeListener(PropertyChangeListener pcl){
        pcs.addPropertyChangeListener(pcl);
    }
    public void removePropertyChangeListener(PropertyChangeListener pcl){
        pcs.removePropertyChangeListener(pcl);
    }
}
```

(3)设计测试程序。

```
package beanTest;
import java.beans.*;
public class TestBindStudent{
    public TestBindStudent(){
    }
    public static void main(String args[]){
        BindStudent st=new BindStudent();
        System.out.println("初始信息:");
        st._toString();
        st.addPropertyChangeListener(new PropertyChangeListener(){
            public void propertyChange(PropertyChangeEvent e){
                System.out.println("University changed:"+e.getNewValue());
            }
        });
        st.setUniversity("BPU");
        System.out.println("最后信息:");
        st._toString();
    }
}
```

(4)程序运行结果如图 4-4 所示。

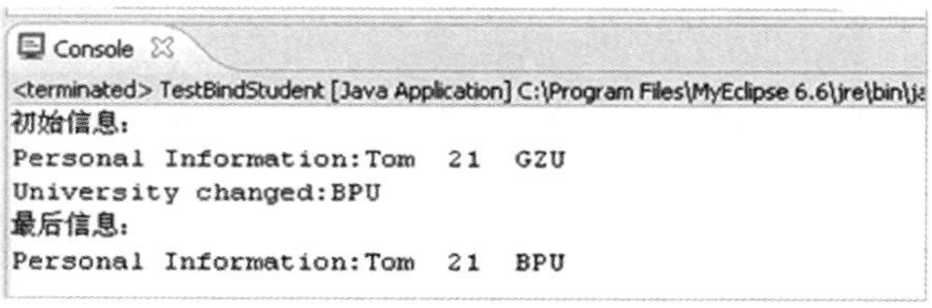

```
Console
<terminated> TestBindStudent [Java Application] C:\Program Files\MyEclipse 6.6\jre\bin\ja
初始信息:
Personal Information:Tom  21  GZU
University changed:BPU
最后信息:
Personal Information:Tom  21  BPU
```

图 4-4　程序运行结果 2

3.设计学生 Bean，将年龄属性设置为约束属性，当年龄小于 20 岁时，不能修改该属性值。

(1)约束属性项目结构如图 4-5 所示。

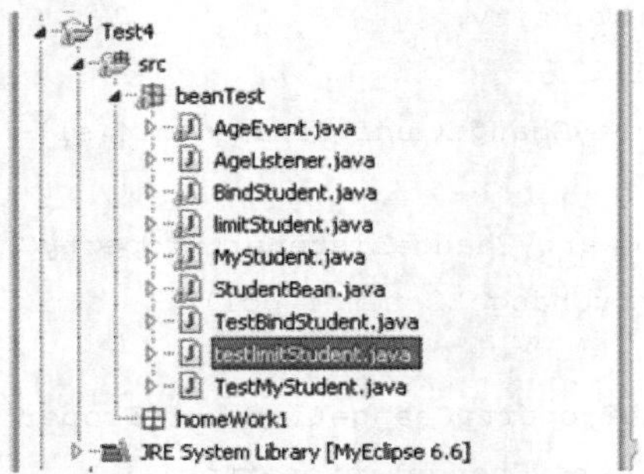

图 4-5　约束属性项目结构图

(2)定义约束属性类 limitStudent，实现 Serializable 接口。

```
package beanTest;
import java.beans.PropertyVetoException;
import java.beans.VetoableChangeListener;
import java.beans.VetoableChangeSupport;
import java.io.Serializable;
public class limitStudent implements Serializable{
    String name;
    int age;
    VetoableChangeSupport vcs;
    String university;
    public limitStudent(){
        vcs= new VetoableChangeSupport(this);
        setName("Tom");
        setAge(21);
        setuniversity("GZU");
    }
    public void_toString(){
        System.out.println(" personal Information:" + name + " " + age +" " +
        university);
    }
    public void setName(String s){
        name=s;
    }
    public String getname(){
        return name;
    }
    public void setAge(int i){
        int old;
```

```
            old=age;
            try{
                vcs.fireVetoableChange("age",old,i);
                age=i;
            }catch(PropertyVetoException e){
             System.out.println(e);
        }
      }
      public int getAge(){
          return age;
      }
      public void setuniversity(String s1){
          university=s1;
      }
      public String getuniversity(){
          return university;
      }
      public void addVetoableChangeListener(VetoableChangeListener vcl){
          vcs.addVetoableChangeListener(vcl);
      }
      public void removePropertyChangeListener(VetoableChangeListener vcl){
           vcs.removeVetoableChangeListener(vcl);
      }
}
```

(3)编写测试程序。

```
package beanTest;
import java.beans.PropertyChangeEvent;
import java.beans.PropertyVetoException;
import java.beans.VetoableChangeListener;
public class testlimitStudent{
    public testlimitStudent(){
    }
    public static void main(String[] args){
        limitStudent st=new limitStudent();
        System.out.println("初始信息:");
        st._toString();
        st.addVetoableChangeListener(new VetoableChangeListener(){
          public void vetoableChange(PropertyChangeEvent e)throws PropertyVetoException{
              if(Integer.parseInt((e.getNewValue()).toString())<=20)
                  throw (new PropertyVetoException("年龄太小"+e.getNewValue(),e));
              System.out.println("university change:"+e.getNewValue());
          }
        });
```

```
        st.setAge(19);
        System.out.println("最后信息:");
        st._toString();
    }
}
```

(4)程序运行结果如图 4-6 所示。

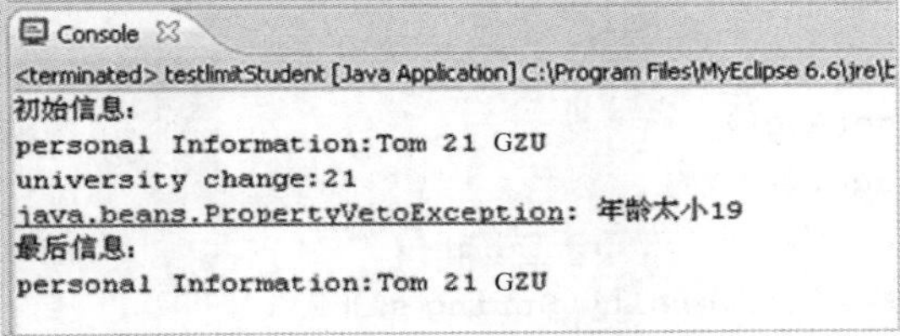

图 4-6　程序运行结果

4.5 实验总结

通过本次实验掌握 JavaBean 组件技术的基本概念;JavaBean 的基本编写规范;JavaBean 的属性、方法和事件;与 JavaBean 技术密切相关的 Java 反射和内省机制。实验实现了 JavaBean 的事件定制、JavaBean 的绑定属性及 JavaBean 的约束属性的设计方法和相关技术。初步了解 JavaBean 的包装和使用技术。初步建立了通过 JavaBean 可以实现软件工程的一个最根本的原则:强内聚、弱耦合,了解 JavaBean 组件技术所要实现的软件设计技术必须遵循的基本思想方法。

4.6 课后思考题

1.一个 JavaBean 有两个属性,即 name、age,按照 JavaBean 规范,这个 JavaBean 对应的方法名应该是什么?

2.一个 JavaBean,类名为 myConnection,根据 JavaBean 规范,这个 JavaBean 必须具有的构造方法形式是什么?

3.JavaBean 属性的类型有哪些?

4.JavaBean 能够体现软件工程的哪些原则?

5.Java 中 Vector 类提供了实现可增长数组的功能,Vector 的三个构造函数是什么?

实验 5

Servlet 创建及使用

5.1 实验目的

1.了解使用 Servlet API
2.理解 Servlet 的生命周期
3.掌握使用 MyEclipse 中开发 Servlet 的方法和步骤
4.熟练掌握 WebLogic 容器下 Servlet 部署和访问的方法

5.2 实验环境

1.MyEclipse 插件平台
2.WebLogic(或 Tomcat)容器
3.MySQL 数据库

5.3 实验知识背景

5.3.1 Servlet 简介

Servlet 是一种基于 Java 技术的、运行在服务器端的 Web 组件，由服务器中的 Servlet 容器所管理。Servlet 是独立于平台的 Java 类，编写一个 Servlet，实际上就是按照 Servlet 的规范来编写一个 Java 类，Servlet 被编译为平台独立的字节码，可以被 Web 服务器加载和执行，类似于 Applet 被浏览器加载和执行。Servlet 从客户端(通过 Web 服务器)接收请求，执行某种作业，然后返回结果。

5.3.2 Servlet 的工作原理

Servlet 能够用于处理客户端的请求，并将处理的请求响应给客户端，Servlet 真正处理客户端请求的阶段是 Servlet 执行阶段。Servlet 采用 Request/Response 模式进行工作，执行原理如图 5-1 所示。

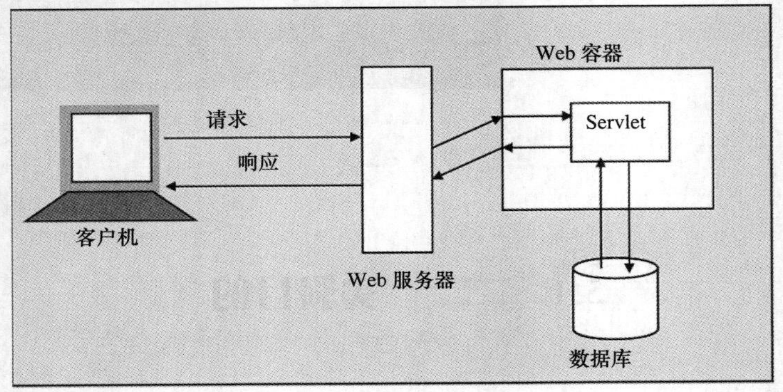

图 5-1　Servlet 的执行原理

1.客户端(一般是 Web 浏览器)将 HTTP 请求发送给 Web 服务器。

2.Web 服务器接收该请求并将其转发给相应的 Servlet。如果该 Servlet 尚未被加载，Web 服务器将把它加载到 Java 虚拟机。

3.Servlet 接收该请求并执行相应处理。

4.Servlet 将响应发送给 Web 服务器。

5.Web 服务器将响应转发给客户端。

5.3.3 Servlet 的应用范围

Servlet 的功能涉及范围广泛，例如：

1.创建并返回一个包含基于客户请求性质的动态内容的完整 HTML 页面。

2.创建可嵌入到现有 HTML 页面中的 HTML 页面。

3.可以同时处理来自客户端的不同请求。

4.与其他服务器资源通过请求的转发(如数据库和其他服务器上的 Servlet 程序等)进行通信。

5.可以定义彼此之间共同工作的激活代理，每个代理者都是一个 Servlet，而且代理者能够在它们之间传送数据。

5.3.4 开发 Servlet 的基本步骤

从开发者的角度来讲，Servlet 是运行在 Web 容器里的一个 Java 类，必须直接或间接实现 javax.servlet.Servlet 接口，并且需要在 web.xml 文件中进行配置，然后把包含该 Servlet 的 Web 应用程序部署到 Web 容器里。概括起来，开发一个 Servlet 大致需要以下几个步骤：

1.编写 Servlet 源码并编译。

2.在 web.xml 文件中进行相应的配置。

3.将包含该 Servlet 的 Web 应用程序部署到 Web 容器里，并启动 Web 容器。

4.通过浏览器访问该 Servlet。

5.3.5 理解 Servlet 的生命周期

Servlet 的生命周期是由 Servlet 容器来控制的，有良好生命周期的定义，包括如何加载、实例化、初始化、处理客户端请求以及如何移除。Servlet 的生命周期由 javax.servlet.Servlet 接口的 init()、service()和 destroy()方法表达，这三个方法必须全部实现。

```
public interface Servlet{
    void init(ServletConfig config);
    void service(ServletRequest request,ServletResponse response);
    void destroy();
}
```

Servlet 的生命周期分为以下 4 个阶段：

1.加载和实例化

Servlet 容器负责加载和实例化，当 Servlet 启动时，或者当 Servlet 容器检测到来自客户端的请求，需要某个特定的 Servlet 进行处理，这时就由服务器加载并实例化 Servlet。Servlet 容器是否在容器启动时自动加载 Servlet，可以在 web.xml 中的<load-on-startup>属性中配置。

在 web.xml 文件中为 Servlet 设置<load-on-startup>元素，代码如下：

```
<servlet>
    <servlet-name>servlet1</servlet-name>
    <servlet-class>servlet.servlet1</servlet-class>
    <load-on-startup>0</load-on-startup>
</servlet>
```

说明：

在 Servlet 的配置当中，“<load-on-startup>0</load-on-startup>”的含义是标记容器是否在启动的时候就加载该 Servlet。当值为 0 或者大于 0 时，表示容器在应用启动时就加载该 Servlet；当值为负数或者没有指定时，则指示容器在该 Servlet 被请求时才加载。正数的值越小，启动该 Servlet 的优先级越高。

2.初始化

在 Servlet 实例化以后，Servlet 容器将调用 init()方法来初始化该对象。在调用 init()方法之前，要确保 Servlet 容器已完成初始化工作(例如：建立数据库连接、通过 ServletConfig 对象获取配置信息)。若初始化失败，将抛出 ServletException 或 UnavailableException 异常，实例销毁。对于每一个 Servlet 实例来说，只初始化一次，GenericServlet 提供了两种形式的 init()方法。

第一种是：

```
public void init(ServletConfig config){
    super.init();//初始化工作
}
```

另一种是：

```
public void init(){ //被 init(ServletConfig config) 调用
    //初始化工作
}
```

从初始化的两种 init()方法的格式来看，重载第二种形式的 init()方法更简便。

3.请求处理

Servlet 实例化后接收客户端请求和做出响应，都是通过调用 service()方法来实现的。每个客户端请求都有其自己的 service()方法，这些方法接收客户端请求，并且返回相应的响应结果。由于 Servlet 采用多线程机制来提供服务，该方法可以被同时、多次地调用。每一个请求都调用自己的 service()方法，但要注意线程安全。

用户在实现具体的 Servlet 时，一般不重载 service()方法，Web 容器在调用 service()方法时，会根据请求方式的不同自动调用 doGet()、doPost()、doPut()和 doDelete()中的一种或几种，因此，只要重载对应的 doXxx()方法即可。例如，重载 doGet()方法，格式如下：

```
public void doGet(HttpServletRequest req,HttpServletResponse resp){
    ......
}
```

重载 doPost()方法，格式如下：

```
public void doPost(HttpServletRequest req,HttpServletResponse resp){
    ......
}
```

4.服务终止

服务器通过调用 destroy()方法释放 Servlet 运行时所占用的资源，该方法不用抛出异常。

```
public void destroy(){
    //释放在 init()中获得的资源，如关闭数据库的连接
    super.destroy();
}
```

在整个 Servlet 的生命周期中，创建 Servlet 实例、调用实例的 init()和 destroy()方法都只进行一次，当初始化完成后，Servlet 容器会将该实例保存在内存中，通过调用它的 service()方法为接收到的请求服务。

5.4 实验内容与步骤

5.4.1 MyEclipse中开发Servlet的步骤

(1)打开 MyEclipse 开发平台,新建一个 Web 项目,如图 5-2 所示。

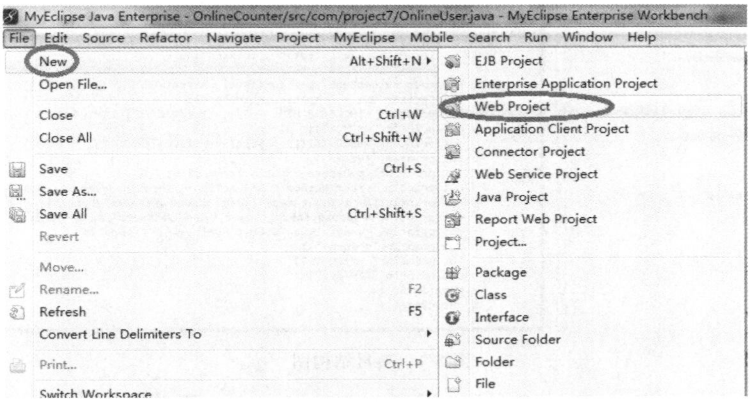

图 5-2 新建 Web 项目

(2)在弹出的对话框中填入项目名称,如图 5-3 所示。

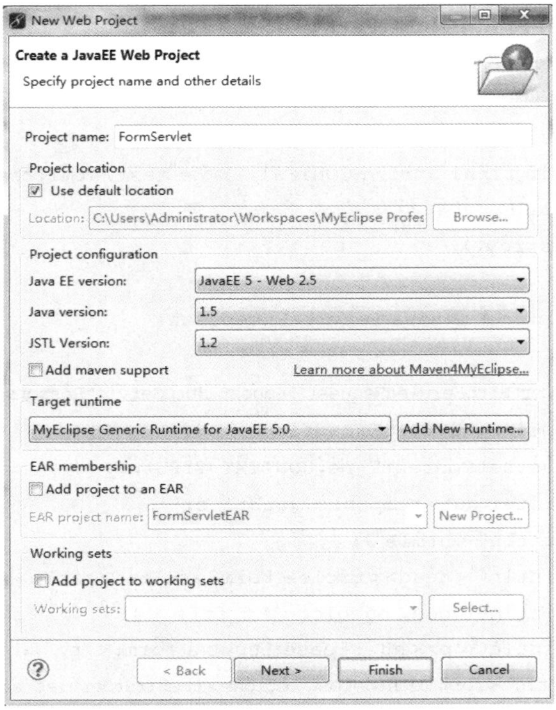

图 5-3 输入 Web 项目名称

(3)设计 Servlet,继承自 HttpServlet。

示例 1 右击选择 FormServlet 项目的 src 目录,建立一个包(如 myServlet)添加 Java 类文件 FormServlet.java。FormServlet 类继承自 HttpServlet 类,因此是自定义的 Servlet。项目结构如图 5-4 所示。

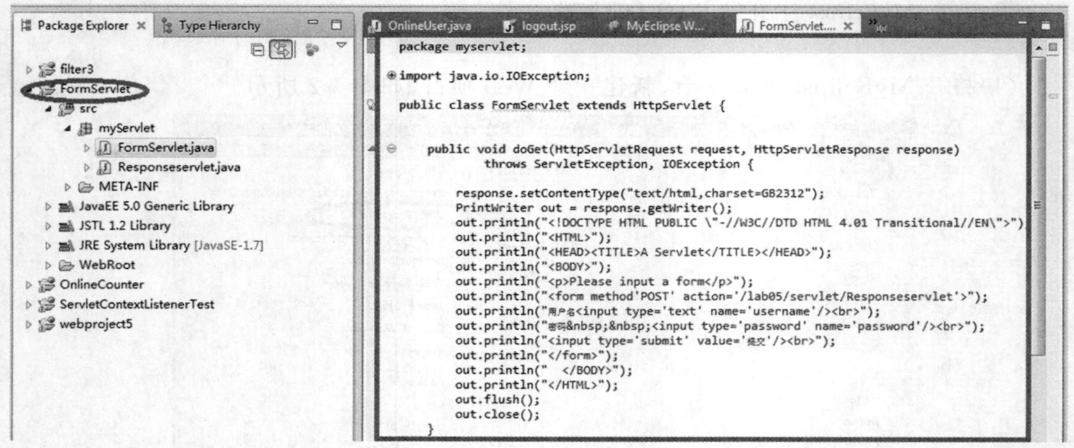

图 5-4 项目结构图

1.设计 servlet,命名为 FormServlet,继承自 HttpServlet,覆盖基类的 destroy()、init()、doGet()和 doPost()方法。

```
package com.myServlet;
import javax.servlet.*;
import javax.servlet.http.*;
import java.io.*;
public class FormServlet extends HttpServlet{
    private static final String CONTEXT_TYPE="text/html;charset=GBK";
    public void destroy(){
        super.destroy();
    }
    public void init() throws ServletException{
    }
    public void doGet(HttpServletRequest request,HttpServletResponse response) throws
          IOException,ServletException{
          response.setContentType(CONTEXT_TYPE);
          PrintWriter out=response.getWriter();
          out.println("<html>");
          out.println("<head><title> FormServlet</title></head>");
          out.println("<body bgcolor='#ffffff'>");
          out.println("<p><h3>Please input a form:</p>");
          out.println("<form method=post action='/theFirstServlet/servlet/FormServlet'");
          out.println("");
          out.println("用户名:<input type=text name='username'/><br> ");
```

```
        out.println("密码:<input type=password name='password'/><br>");
        out.println("<input type=submit value='提交查询内容'/>");
        out.println("</form>");
        out.println("</body>");
        out.println("</html>");
        out.flush();
        out.close();
    }
    public void doPost(HttpServletRequest request,HttpServletResponse response) throws
      IOException,ServletException{
        doGet(request,response);
    }
}
```

2.编辑 Web 描述文件 web.xml。

```
<?xml version="1.0" encoding="UTF-8"?>
<web-app version="2.5"
    xmlns="http://java.sun.com/xml/ns/javaee"
    xmlns:xsi="http://www.w3.org/2001/XMLSchema-instance"
    xsi:schemaLocation="http://java.sun.com/xml/ns/javaee
    http://java.sun.com/xml/ns/javaee/web-app_2_5.xsd">
<servlet>
    <description>This is the description of my J2EE component</description>
    <display-name>This is the display name of my J2EE component</display-name>
    <servlet-name>myFirstServlet</servlet-name>
    <servlet-class>com.myServlet.FormServlet</servlet-class>
</servlet>
<servlet-mapping>
    <servlet-name>myFirstServlet</servlet-name>
    <url-pattern>/servlet-FormServlet</url-pattern>
</servlet-mapping>
<welcome-file-list>
      <welcome-file> index.jsp</welcome-file>
</welcome-file-list>
</web-app>
```

3.部署项目到 WebLogic 中,并启动 WebLogic 服务器。如图 5-5 所示。

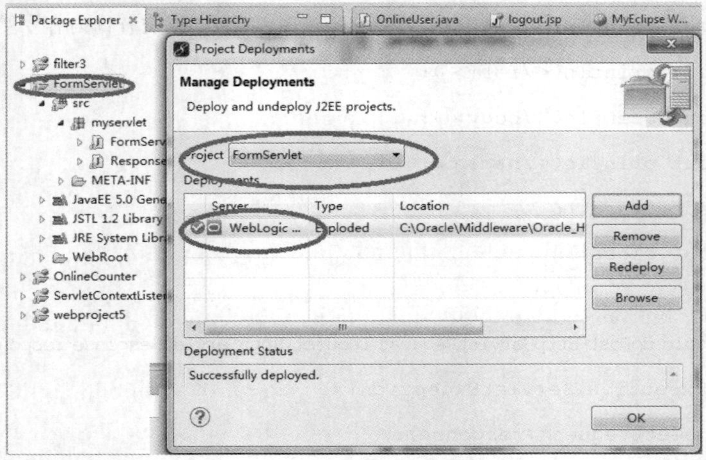

图 5-5　部署项目

4.启动服务器,如图 5-6 所示。

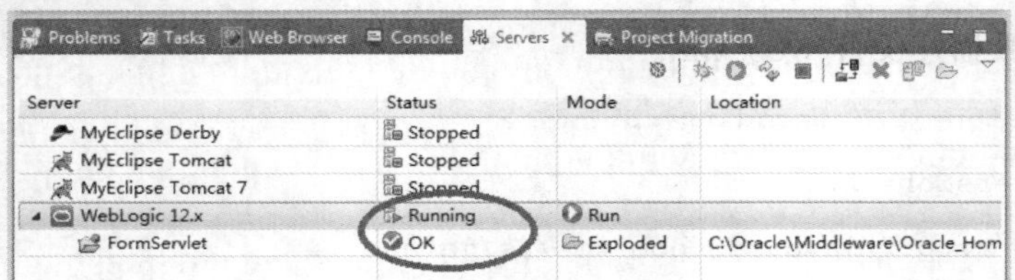

图 5-6　启动服务器

5.在 WebLogic Console 观察项目部署状态,如图 5-7 所示。

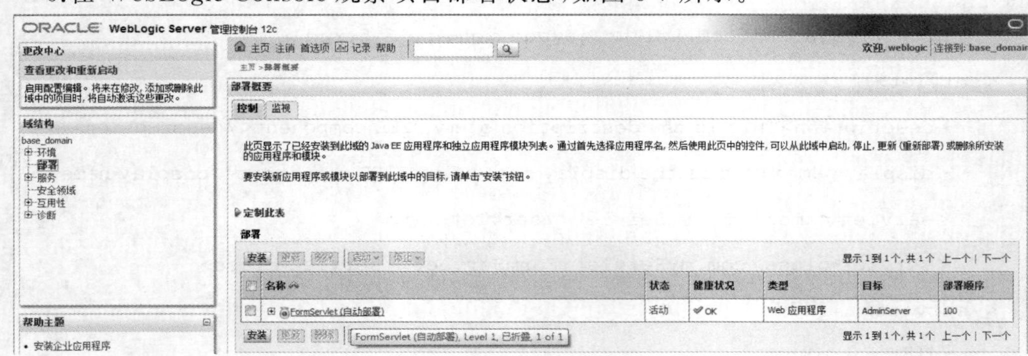

图 5-7　项目部署状态

6.程序运行结果如图 5-8 所示。

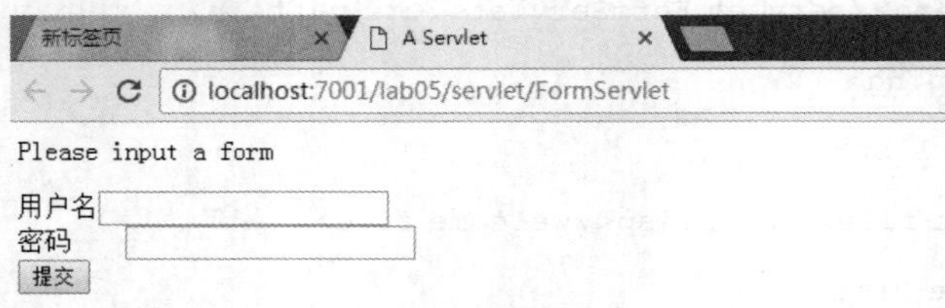

图 5-8　程序运行结果

示例 2 编写一个 StudentServlet，完成下列要求：

1.该 Servlet 提供学生学号查询界面，如图 5-9 所示。

http://localhost:7001/HelloWebModule/studentservlet

学生学号 Input number 查询

图 5-9 学生学号查询界面

2.当学生输入学号，且按下【查询】按钮后，该 Servlet 完成数据库查询，如果存在该学号，将该学生的成绩记录显示输出，如图 5-10 所示。

图 5-10 学号存在时的查询结果

3.如果数据库没有当前用户输入的学生学号，结果将显示无记录，如图 5-11 所示。

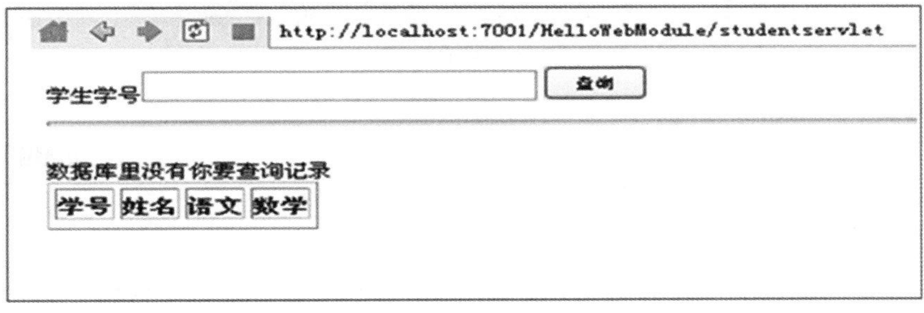

图 5-11 学号不存在时的查询结果

4.该 Servlet 通过 JNDI 访问数据源，进行数据库连接查询。

5.设计后台数据库：学生及考试信息。

学生学号(PK)＋姓名＋语文成绩＋数学成绩。

注意：只编写一个 Servlet 完成上述要求。

6.建立数据库和数据库表,如图 5-12 所示。

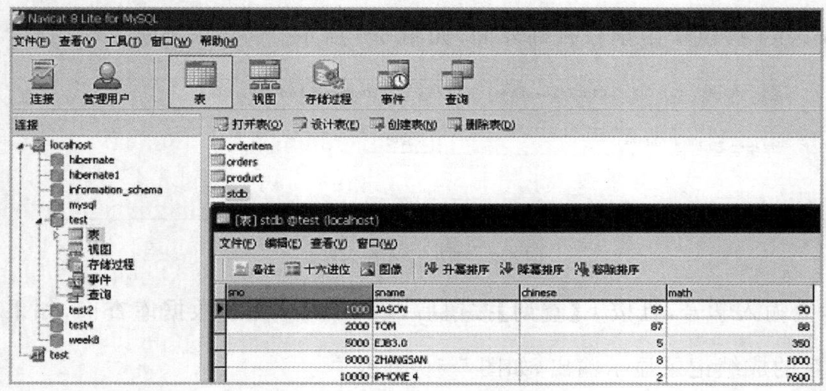

图 5-12　数据库及数据库表结构

7.登录 WebLogic Console 进行 JDBC 数据源设置,如图 5-13 所示。

图 5-13　登录 WebLogic Console

8.设置 JNDI 数据源,如图 5-14 所示。

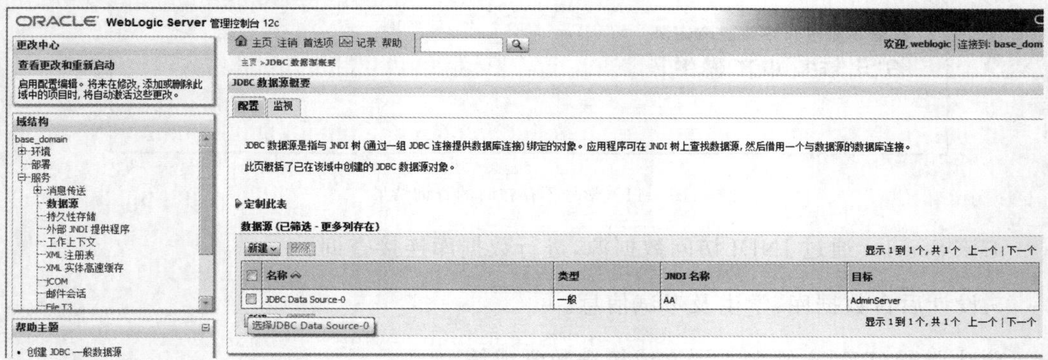

图 5-14　设置 JNDI 数据源

9.设计 servlet，命名为 StudentServlet，继承自 HttpServlet，覆盖基类的 destroy()、init()、doGet()和 doPost()方法。

```
package myservlet;
import java.io.IOException;
import java.io.PrintWriter;
import java.sql.Connection;
import java.sql.PreparedStatement;
import java.sql.ResultSet;
import java.sql.Statement;
import java.util.Properties;
import java.util.Vector;
import javax.naming.Context;
import javax.naming.InitialContext;
import javax.servlet.ServletException;
import javax.servlet.http.HttpServlet;
import javax.servlet.http.HttpServletRequest;
import javax.servlet.http.HttpServletResponse;
import javax.sql.DataSource;
public class StudentServlet extends HttpServlet{
    public StudentServlet(){
        super();
    }
    public void destroy(){
        super.destroy();// Just puts "destroy" string in log
        // Put your code here
    }
    private static final String Content_Type="text/html;charset=GB2312";
    public void doGet(HttpServletRequest request,HttpServletResponse response)
        throws ServletException,IOException{
        response.setContentType(Content_Type);
        PrintWriter out=response.getWriter();
        out.println("<HTML>");
        out.println("<HEAD><TITLE>StudentServlet</TITLE></HEAD>");
        out.println("<BODY bgcolor=\"#ffffff\">");
        if((request.getParameter("sno"))==null
                ||(request.getParameter("sno").equals("input number"))){
            out.println("<form method='post' action='/Lab5/servlet/StudentServlet'>");
            out.println("");
            out.println("学号查询:<input type='text' name='sno' value='input number'/>");
            out.println("<input type='submit' value='查询' /><br>");
```

```
            out.println("</form>");
        }
        else if(!(request.getParameter("sno").equals("input number"))){
            Vector vc=new Vector();
            out.println("<form method='post' action='/Lab5/servlet/StudentServlet'>");
            out.println("");
            out.println("学生学号:<input type='text' name='sno'/>");
            out.println("<input type='submit' value='查询'/><br>");
            out.println("</form>");
            out.println("<hr>");
            out.println("<table border='1'><tr>");
            out.println("<th>学号</th><th>姓名</th><th>语文</th><th>数学</
            th></tr>");
            try {
                if((vc=accessDB(Integer.parseInt((String) request
                        .getParameter("sno"))))!=null){
                    out.println("<tr>");
                    out.println("<td>"+ vc.elementAt(0)+ "</td>");
                    out.println("<td>"+ vc.elementAt(1)+ "</td>");
                    out.println("<td>"+ vc.elementAt(2)+ "</td>");
                    out.println("<td>"+ vc.elementAt(3)+ "</td>");
                    out.println("</tr>");
                }
            }catch (Exception e){
                out.println("<tr>");
                out.println("<td>没有记录</td>");
                out.println("</tr>");
                out.println("数据库里没有你要查询记录");
                e.printStackTrace();
              }
            out.println("</table>");
        }
        out.println("</BODY>");
        out.println("</HTML>");
        out.flush();
        out.close();
    }
    public void doPost(HttpServletRequest request,HttpServletResponse response)
            throws ServletException,IOException{
        doGet(request,response);
    }
    public void init() throws ServletException{
```

```
    // Put your code here
}
//通过 WebLogic 的 JNDI 服务来查询数据库数据
public Vector accessDB(int id){
    Vector vc=new Vector();
    DataSource ds=null;
    Context ctx;
    Connection myConn=null;
    final String JNDI_DATABASE_NAME="AA";
    Properties ht=new Properties();
    ht.put(Context.INITIAL_CONTEXT_FACTORY,
            "weblogic.jndi.WLInitialContextFactory");
    ht.put(Context.PROVIDER_URL,"t3://localhost:7001");
    try{
        ctx=new InitialContext(ht);
        ds=(javax.sql.DataSource) ctx.lookup(JNDI_DATABASE_NAME);
    }catch(Exception e){
        e.printStackTrace();
    }
    if(ds==null){
        System.out.println("Eorror !");
    }
    else{
        System.out.println("Connection is OK !");
    }
    PreparedStatement myStatement=null;
    ResultSet mySet=null;
    try{
        myConn=ds.getConnection();
        myStatement=myConn.prepareStatement("select* from stdb where sno=?");
        myStatement.setInt(1,id);
        mySet=myStatement.executeQuery();
        while(mySet.next()){
            vc.add(Integer.toString(mySet.getInt("sno")));
            vc.add(mySet.getString("sname"));
            vc.add(Integer.toString(mySet.getInt("chinese")));
            vc.add(Integer.toString(mySet.getInt("math")));
        }
        myStatement.close();
        mySet.close();
        myConn.close();
        return vc;
```

```
        }catch (Exception e){
            e.printStackTrace();
        }
        return vc;
    }
    public static void main(String args[]){
        StudentServlet aa=new StudentServlet();
        System.out.println(aa.accessDB(100).elementAt(1));
    }
}
```

10.修改项目描述文件 web.xml。

```
<?xml version="1.0" encoding="UTF-8"?>
<web-app version="2.5"
    xmlns="http://java.sun.com/xml/ns/javaee"
    xmlns:xsi="http://www.w3.org/2001/XMLSchema-instance"
    xsi:schemaLocation="http://java.sun.com/xml/ns/javaee
    http://java.sun.com/xml/ns/javaee/web-app_2_5.xsd">
  <display-name></display-name>
  <servlet>
    <description>This is the description of my J2EE component</description>
    <display-name>This is the display name of my J2EE component</display-name>
    <servlet-name>StudentServlet</servlet-name>
    <servlet-class>myservlet.StudentServlet</servlet-class>
  </servlet>
  <servlet-mapping>
    <servlet-name>StudentServlet</servlet-name>
    <url-pattern>/servlet/StudentServlet</url-pattern>
  </servlet-mapping>
  <welcome-file-list>
    <welcome-file>index.jsp</welcome-file>
  </welcome-file-list>
</web-app>
```

11.程序运行结果如图 5-15 所示。

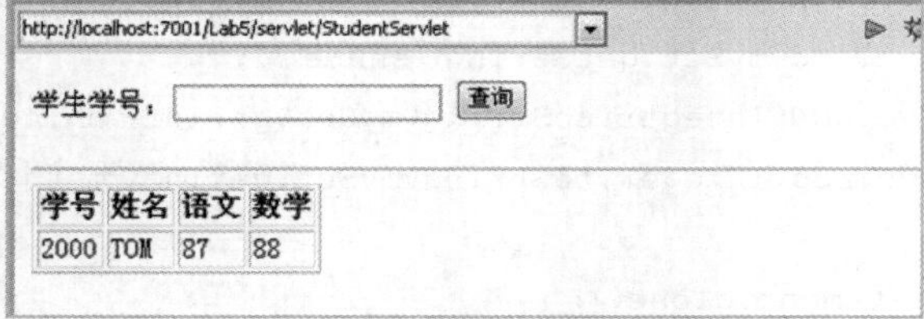

图 5-15　程序运行结果

5.5 实验总结

本实验使用 HTTP 传输协议，运用 Servlet API 中 javax.servlet 和 javax.servlet.http 两个包，按照 Servlet 规范编写 Servlet，用于在客户端发送 HTTP 请求及接收响应结果和服务器接收 HTTP 请求并返回响应。Servlet 既可以扩展 GenericServlet 类，也可以扩展 HttpServlet 类。Servlet 的生命周期包含三个方法，分别是 init()、service()和 destroy()。有三种类型的 Servlet，即标准 Servlet、监听器 Servlet 和过滤器 Servlet。MyEclipse 是用于开发 Web 应用程序(如 Servlet、Applet 和 JSP)的一种非常有用的软件，在 MyEclipe 中开发 Servlet 的基本步骤是：创建 Web 项目、编写 Servlet、在 web.xml 中配置 Servlet、部署项目到容器、启动容器服务器、打开 Web 浏览器访问 Servlet。

5.6 课后思考题

1.什么是 Servlet，有哪些特点？

2.Servlet 的生命周期包括哪几个阶段，每个阶段实现的过程是什么？

3.如何在 web.xml 文件中配置 Servlet？

4.Request 对象的主要方法有哪些？

5.Servlet 执行时一般实现哪几个方法？

6.Servlet 创建的基本步骤有哪些？

实验 6 Servlet 会话跟踪

6.1 实验目的

1.掌握创建 Servlet 会话的方法
2.熟练掌握 Servlet 会话跟踪技术
3.理解 Servlet 通信方法
4.掌握创建 Servlet 程序以访问上下文状态的方法
5.了解开发 Servlet 程序以修改上下文状态的步骤

6.2 实验环境

1.MyEclipse 插件平台
2.WebLogic(或 Tomcat)容器

6.3 实验知识背景

在 Web 应用中,有必要保持同一客户的会话,进行会话跟踪。Internet 协议提供两种类型保持客户端状态:有状态(Telnet、FTP)和无状态(HTTP)。服务器一旦响应完客户端的请求之后,便断开与其之间网络连接,该客户端下次请求将重新建立网络连接,服务器判断是否为同一个客户端发出请求。因为 HTTP 协议是无状态的,同一客户端所发出的每次请求对容器而言都是一个新的客户端,客户端需要一个唯一的会话 ID,即对客户端的第一次请求,容器会生成一个唯一的会话 ID,并通过响应将其返回给客户端;客户端在以后的每一个请求中发回这个会话 ID,容器看到 ID 后,就会找匹配的会话,并把这个会话与请求关联,这些依赖于会话跟踪和 Servlet 通信技术。

6.3.1 会话管理

对于一个网上商城的网站来说,用户通常浏览各种各样的商品,然后将这些商品放入购物车,最后到结算中心进行结算。在这个过程中,服务器端是如何得知这些商品是同一用户选购的呢？如何来跟踪这位用户呢？这就是会话(Session)的概念,记录在一定的时间段内,来自同一客户端的一系列HTTP请求,服务器端正是利用会话机制来跟踪用户的。图6-1是三种会话跟踪技术。

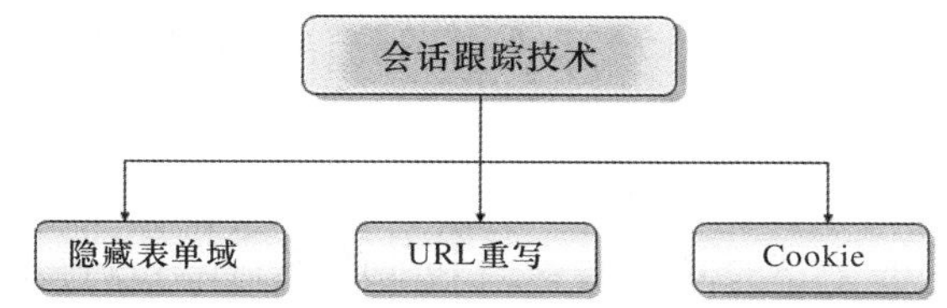

图6-1 会话跟踪技术

会话技术是基于Cookie的,可以通过Cookie将信息保存到客户端。Cookie用名-值对的形式来保存用户的一些信息,如用户名、密码等。但是Cookie有个缺陷,如果用户选择多个产品,就需要产生多个Cookie来保存,并且Cookie只能保存String类型的数据,不能保存对象。这时可以采用会话,它的基本思路是:给每一个客户端分配一个不重复的ID,用来区分不同的客户端,将对应客户端需要保存的数据保存在服务器的内存中。

在Servlet中,Session封装在javax.servlet.http.HttpSession接口中,使用Session可以分为如下三步:

(1)获取Session,通过HttpRequest对象调用getSession()或getSession(boolean value)方法。

(2)存储数据到Session或从Session读取数据,通过以下方法:

①getAttribute(String name):查找以前存储的值,如果不存在则返回null。在使用此方法返回对象前,一定要检查它是否为null,且要保证其类型为Object,通常要强制类型转换。

②setAttribute(String name,Object object):设置属性。

③removeAttribute(String name):移除属性。

(3)最后销毁Session:可以使Session过期,或调用invalidate()强制其失效。

6.3.2 Servlet通信

Web开发过程中,各种资源之间需要进行通信完成一定的功能。Servlet的通信包括与浏览器之间的通信、Servlet与Servlet、Servlet与JSP等组件与组件之间的通信。本节将学习Servlet通信过程中的一些技巧。Servlet之间通信依赖于ServletContext上下文。在Web应用中,Servlet之间可以相互调用及共享信息(一个JVM中只能有一个ServletContext)。

6.3.3 请求转发

用户向服务器端发出了一次HTTP请求，该请求可能会经过多个信息资源处理以后才返回给用户，各个信息资源使用请求转发机制相互转发请求，但是用户是感觉不到请求转发的。实际的Web开发过程中，会使用大量请求转发机制。Servlet使用委托机制——RequestDispatcher(请求分派)接口的forward()和include()方法实现多个Servlet信息共享。

RequestDispatcher接口有两种方法，在Servlet API讲解部分介绍过这两种方法，一种是forward，用于将请求从一个Servlet传递到服务器上的另外的Servlet、JSP页面或者是HTML文件。另外一种是include，该方法用于在响应中包含其他资源的内容。两者的区别是：利用include方法将请求转发给其他的Servlet，被调用的Servlet对该请求做出的响应将并入原先的响应对象中，原先的Servlet还可以继续输出响应信息；而利用forward方法，则是将请求转发给其他的Servlet，将由被调用的Servlet负责对请求做出响应，而原先Servlet的执行则终止。

6.4 实验内容与步骤

1.实验要求：使用Servlet编写聊天程序完成如下功能：

(1)用户输入昵称后，可以参加聊天，该昵称为使用聊天应用程序的用户名。

(2)用户在文本框中输入发送的消息，单击命令按钮参与聊天。

(3)用户可以看见所有参与聊天的人和全部聊天记录，没有其他的权限控制。

(4)用户可以使用中文聊天。

2.UserServlet生成用户界面，接收用户输入的昵称，如图6-2所示。

图6-2 聊天系统主界面

3.参考代码。

(1)项目结构如图6-3所示。

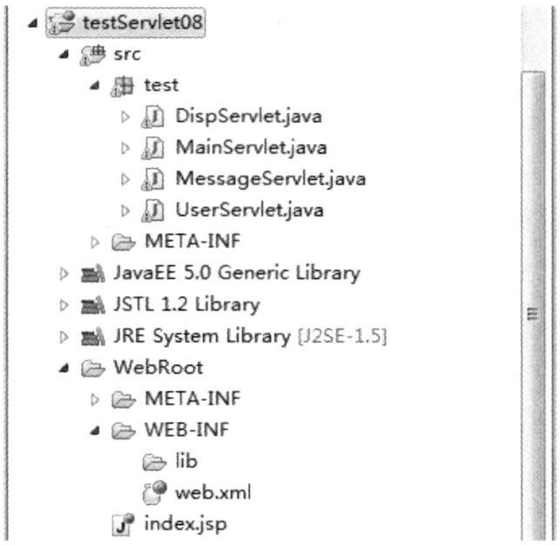

```
testServlet08
  src
    test
      DispServlet.java
      MainServlet.java
      MessageServlet.java
      UserServlet.java
    META-INF
  JavaEE 5.0 Generic Library
  JSTL 1.2 Library
  JRE System Library [J2SE-1.5]
  WebRoot
    META-INF
    WEB-INF
      lib
      web.xml
    index.jsp
```

图 6-3　项目结构图

(2)设计 UserServlet,继承自 HttpServlet,作为聊天系统的入口。客户端访问该 Servlet,输入聊天用户名(昵称),并链接到聊天主界面 MainServlet。

```
package test;
import java.io.IOException;
import java.io.PrintWriter;
import java.util.Vector;
import javax.servlet.ServletContext;
import javax.servlet.ServletException;
import javax.servlet.http.HttpServlet;
import javax.servlet.http.HttpServletRequest;
import javax.servlet.http.HttpServletResponse;
public class UserServlet extends HttpServlet{
    public void doGet(HttpServletRequest request,HttpServletResponse response)
            throws ServletException,IOException{
        response.setContentType("text/html;charset=GBK");
        request.setCharacterEncoding("GBK");
        PrintWriter out=response.getWriter();
        out.println("<!DOCTYPE HTML PUBLIC \"-//W3C//DTD HTML 4.01 Transitional//EN\">");
        out.println("<HTML>");
        out.println("<HEAD><TITLE>A Servlet</TITLE></HEAD>");
        out.println("<BODY>");
        out.println("<h3>进入聊天系统....");
        ServletContext context=getServletContext();
        if(context.getAttribute("userlist")==null){
            Vector aduser=new Vector();
            context.setAttribute("userlist",aduser);
        }
        if(context.getAttribute("messagelist")==null){
            Vector message=new Vector();
```

```
            context.setAttribute("messagelist",message);
        }
        out.println("<form method='post' action=
        '/testServlet08/servlet/MainServlet'>");
        out.println("昵称<input type='text' name='uname'/>");
        out.println("<input type=submit value='登录'>");
        out.println("</form>");
        out.println("</BODY>");
        out.println("</HTML>");
        out.flush();
        out.close();
    }
    public void doPost(HttpServletRequest request,HttpServletResponse response)
            throws ServletException,IOException{
        doGet(request,response);
    }
}
```

(3)MainServlet 读出 Servlet 上下文中保存的用户列表向量，如果向量为 null，则新建一个向量，并在用户列表向量中添加该用户。设计 MainServlet，将任务分派给 MessageServlet。

```
package test;
import java.io.IOException;
import java.io.PrintWriter;
import java.util.Vector;
import javax.servlet.RequestDispatcher;
import javax.servlet.ServletContext;
import javax.servlet.ServletException;
import javax.servlet.http.HttpServlet;
import javax.servlet.http.HttpServletRequest;
import javax.servlet.http.HttpServletResponse;
public class MainServlet extends HttpServlet{
    public void doGet(HttpServletRequest request,HttpServletResponse response)
            throws ServletException,IOException{
        response.setContentType("text/html;charset=GBK");
        request.setCharacterEncoding("GBK");
        PrintWriter out=response.getWriter();
        out.println("<!DOCTYPE HTML PUBLIC \"-//W3C//DTD HTML 4.01 Transitional//EN\">");
        out.println("<HTML>");
        out.println("<HEAD><TITLE>A Servlet</TITLE></HEAD>");
        out.println("<BODY>");
        out.println("<h3>进入聊天系统...");
```

```
        ServletContext context=getServletContext();
        Vector vc=(Vector)context.getAttribute("userlist");
        String username=request.getParameter("uname");
        vc.add(username);
        RequestDispatcher view=request.getRequestDispatcher
        ("/servlet/MessageServlet? uname="+username);
        view.forward(request,response);
        out.println("</BODY>");
        out.println("</HTML>");
        out.flush();
        out.close();
    }
    public void doPost(HttpServletRequest request,HttpServletResponse response)
            throws ServletException,IOException{
        doGet(request,response);
    }
}
```

(4)MessageServlet 显示一个文本框和一个命令按钮,在文本框中输入要发送的消息。使用 requestsetCharacterEncoding("GBK")处理中文。使用 iframe 将 DispServlet 添加到 MessageServlet 中。

```
package test;
import java.io.IOException;
import java.io.PrintWriter;
import java.util.Vector;
import javax.servlet.ServletContext;
import javax.servlet.ServletException;
import javax.servlet.http.HttpServlet;
import javax.servlet.http.HttpServletRequest;
import javax.servlet.http.HttpServletResponse;
public class MessageServlet extends HttpServlet{
    public void doGet(HttpServletRequest request,HttpServletResponse response)
            throws ServletException,IOException{
        response.setContentType("text/html;charset=GBK");
        request.setCharacterEncoding("GBK");
        PrintWriter out=response.getWriter();
        out.println("<! DOCTYPE HTML PUBLIC \"-//W3C//DTD HTML 4.01 Transitional//EN\">");
        out.println("<HTML>");
        out.println("<HEAD><TITLE>A Servlet</TITLE></HEAD>");
        out.println("<BODY>");
        ServletContext context=getServletContext();
        String usr=request.getParameter("uname");
```

```
        String str=request.getParameter("message");
        Vector hsh=(Vector)context.getAttribute("messagelist");
        if(str!=null){
            hsh.add(usr+":"+str);
        }
        out.print("<iframe frameboder=0 heigth=146 marginheight=0 marginwidth=0"+
        " scrolling = auto src ='/testServlet08/servlet/DispServlet' width =' 100%'
        heigth='80%'>");
        out.println("</iframe>");
        out.println("<form method='post' action='/testServlet08/servlet
        /MessageServlet? uname="+ usr+ "'>");
        out.println("<hr>");
        out.println("输入消息:<input type='text' name='message'/>");
        out.println("<input type=submit value='发送'/>");
        out.println("</form>");
        out.println("</BODY>");
        out.println("</HTML>");
        out.flush();
        out.close();
    }
    public void doPost(HttpServletRequest request,HttpServletResponse response)
            throws ServletException,IOException{
        doGet(request,response);
    }
}
```

(5)设计 DisplayServlet 显示所有的聊天消息。用 messagelist 和 userlist 分别显示所有的聊天内容和用户列表。

```
package test;
import java.io.IOException;
import java.io.PrintWriter;
import java.util.Vector;
import javax.servlet.ServletContext;
import javax.servlet.ServletException;
import javax.servlet.http.HttpServlet;
import javax.servlet.http.HttpServletRequest;
import javax.servlet.http.HttpServletResponse;
public class DispServlet extends HttpServlet{
    public void doGet(HttpServletRequest request,HttpServletResponse response)
            throws ServletException,IOException{
        response.setContentType("text/html;charset=GBK");
```

```
        request.setCharacterEncoding("GBK");
        PrintWriter out=response.getWriter();
        ServletContext context=getServletContext();
        out.println("<!DOCTYPE HTML PUBLIC \"-//W3C//DTD HTML 4.01 Transitional//EN\">");
        out.println("<HTML>");
        out.println("<HEAD><TITLE>A Servlet</TITLE></HEAD>");
        out.println("<BODY>");
        out.println("内容");
        Vector hab=(Vector)context.getAttribute("messagelist");
        out.println("<textarea name='info' rows='9'>");
        for(int i=0;i<hab.size();i++){
            out.println(hab.get(i));
        }
        out.println("</textarea>");
        out.println("用户列表");
        out.println("<textarea name='user' rows='9'>");
        Vector vc=(Vector)context.getAttribute("userlist");
        for(int i=0;i<vc.size();i++){
            if(vc.get(i)!=null){
                out.println(vc.get(i));
            }
        }
        out.println("</textarea>");
        out.println("</BODY>");
        out.println("</HTML>");
        out.flush();
        out.close();
    }
    public void doPost(HttpServletRequest request,HttpServletResponse response)
            throws ServletException,IOException{
        doGet(request,response);
    }
}
```

(6)配置 web.xml 文件。

```
<?xml version="1.0" encoding="UTF-8"?>
<web-app version="2.5"
    xmlns="http://java.sun.com/xml/ns/javaee"
    xmlns:xsi="http://www.w3.org/2001/XMLSchema-instance"
    xsi:schemaLocation="http://java.sun.com/xml/ns/javaee
    http://java.sun.com/xml/ns/javaee/web-app_2_5.xsd">
  <servlet>
```

```
    <description>This is the description</description>
    <display-name>This is the display</display-name>
    <servlet-name>UserServlet</servlet-name>
    <servlet-class>test.UserServlet</servlet-class>
  </servlet>
  <servlet>
    <description>This is the description </description>
    <display-name>This is the display </display-name>
    <servlet-name>MainServlet</servlet-name>
    <servlet-class>test.MainServlet</servlet-class>
  </servlet>
  <servlet>
    <description>This is the description</description>
    <display-name>This is the display</display-name>
    <servlet-name>MessageServlet</servlet-name>
    <servlet-class>test.MessageServlet</servlet-class>
  </servlet>
  <servlet>
    <description>This is my J2EE component</description>
    <display-name>This is the display </display-name>
    <servlet-name>DispServlet</servlet-name>
    <servlet-class>test.DispServlet</servlet-class>
  </servlet>
  <servlet-mapping>
    <servlet-name>UserServlet</servlet-name>
    <url-pattern>/servlet/UserServlet</url-pattern>
  </servlet-mapping>
  <servlet-mapping>
    <servlet-name>MainServlet</servlet-name>
    <url-pattern>/servlet/MainServlet</url-pattern>
  </servlet-mapping>
  <servlet-mapping>
    <servlet-name>MessageServlet</servlet-name>
    <url-pattern>/servlet/MessageServlet</url-pattern>
  </servlet-mapping>
  <servlet-mapping>
    <servlet-name>DispServlet</servlet-name>
    <url-pattern>/servlet/DispServlet</url-pattern>
  </servlet-mapping>
  <welcome-file-list>
    <welcome-file>index.jsp</welcome-file>
  </welcome-file-list>
```

</web-app>

(7)程序运行结果如图 6-4 所示。

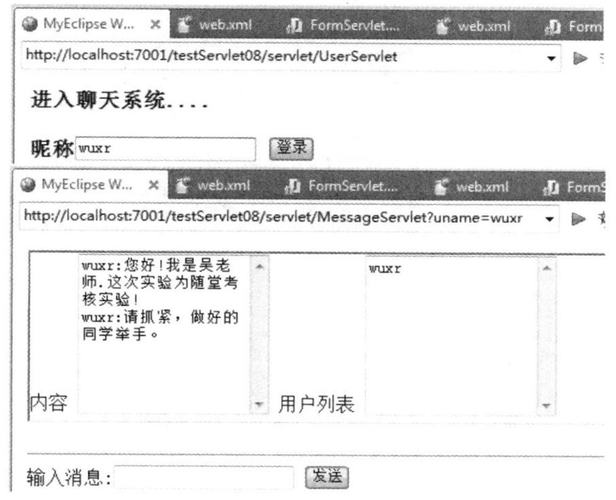

图 6-4　聊天系统程序运行结果

6.5 实验总结

本实验从会话对象的创建、会话跟踪入手，应用 Servlet 上下文状态、Servlet 通信方法等知识，设计创建程序以访问、修改上下文状态。Servlet 能够通过 request 对象获取客户端的请求信息，并能够访问 Session 中的信息，还能对 Cookie 进行操作。此外，Servlet 可以通过 Response 对客户端进行响应，并能够把请求转发给其他的文件处理。

6.6 课后思考题

1.如何创建会话，怎样定义会话跟踪?

2.Servlet 会话跟踪技术有哪些?

3.简述 Servlet 通信方法。

4.试描述 Servlet 上下文状态。

5.简述创建一个访问上下文状态程序的步骤。

实验 7 Servlet 线程安全及过滤器

7.1 实验目的

1.了解 Servlet 的类层次结构
2.掌握 Servlet 的线程安全技术
3.掌握描述 Servlet 过滤器的方法
4.理解过滤器生命周期
5.能够实现简单的过滤器程序设计

7.2 实验环境

1.MyEclipse 插件平台
2.WebLogic(或 Tomcat)容器

7.3 实验知识背景

7.3.1 Servlet API

J2EE 平台提供两个包支撑 Servlet 体系结构：

(1)javax.servlet 包：提供所有 Servlet 类实现的基本接口和继承的基本类。

(2)javax.servlet.http 包：提供编写基于 HTTP 协议的 Servlet 所需的基类。

7.3.2 Servlet 类层次结构

Servlet 类层次结构如图 7-1 所示。

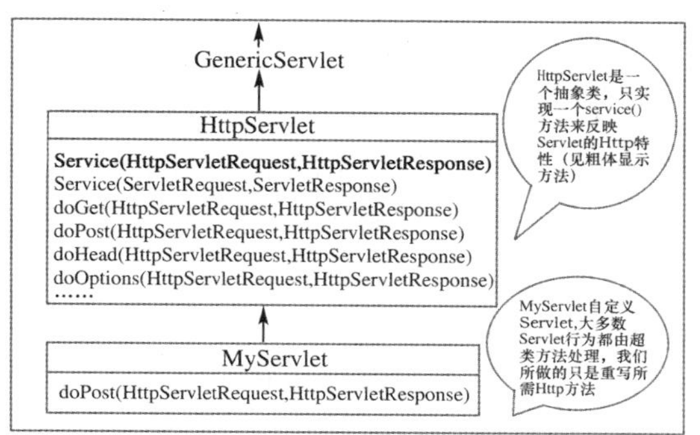

图 7-1 Servlet 类层次结构

1.说明 1

(1)创建协议无关的 Servlet:继承 javax.servlet.GenericServlet 类。

(2)创建基于 HTTP 协议的 Servlet:继承 javax.servlet.http.HttpServlet 类。

(3)GenericServlet 类的 service()方法是抽象方法,它的参数分别为 ServletRequest 和 ServletResponse,前者包含了从客户端发送给服务器的信息,后者则包含了服务器返回给客户端的信息。

(4)HttpServlet 类已经实现了 service()方法,不要在自己的 Servlet 中重写此方法,该方法根据客户端请求方式调用对应的 doXxx()方法,主要是调用 doGet()和 doPost()(默认为 doGet()),这两个方法的参数为 HttpServletRequest 和 HttpServletResponse 对象。

2.说明 2

在用户的继承类中,service()方法会根据请求方法的类型,调用相应的 doXxx(HttpServletRequest req, HttpServletResponse resp)方法。因此,我们在编写 HttpServlet 的派生类时,通常不是覆盖 service()方法,而是重写相应的 doXxx()方法。

3.HttpServletRequest 接口的部分常用方法

HttpServletRequest 接口的部分常用方法,如表 7-1 所示。

表 7-1 HttpServletRequest 接口的部分常用方法

方 法	功能说明
setCharacterEncoding(String env)	设置指定的字符集编码
getParameter(String name)	返回指定参数名对应的数值
getMethod()	返回请求方法的名称
getQueryString()	返回客户端的查询字符串
getRemoteAddr()	返回客户端的 IP 地址
getServletPath()	返回以"/"开头的 Servlet 名称或路径
getCookies()	返回客户端发送的 Cookie

（续表）

方　法	功能说明
getSession()	返回与客户端关联的 HttpSession 对象
getHeader(String name)	返回指定名字的请求头的值
setCharacterEncoding(String env)	设置指定的字符集编码
setContentType(String type)	设置响应体中 MIME 类型及字符集
getWriter()	返回请求体中带缓冲的字符输出流
addCookie(Cookie cookie)	增加 Cookie 到响应中
setHeader(String name,String value)	设置相应的响应头值
addHeader(String name,String value)	把给定的名字和值增加到响应中
setStatus(int sc)	设置响应体的状态码
sendError(int sc,String msg)	向客户端发送的出错信息
sendRedirect(String location)	用给定的 URL 作为重定向地址发送给客户端

7.3.3 Servlet 的配置和初始化参数

1.容器初始化一个 Servlet 时，会为其建一个唯一的 ServletConfig：

(1)容器从 web.xml 中获取 Servlet 初始化参数，并把这些参数交给 ServletConfig。

(2)容器创建 ServletContext 对象并将其存储到 ServletConfig 对象中。

(3)容器调用 Servlet 的 init(ServletConfig)方法并将该 ServletConfig 对象传递给 Servlet。

2.一旦 Servlet 获得 ServletConfig 对象，将完成以下任务：

(1)检索 Servlet 的初始化参数。

(2)访问 ServletContext 对象。

(3)检索 Servlet 名称。

7.3.4 过滤器简介

Servlet 过滤器是一种特殊的 Servlet，可以用来对请求进行过滤。多个过滤器可以形成过滤链。

在请求发送到 Servlet 之前，可以用过滤器截获和处理请求，另外 Servlet 结束工作之后，在响应发回给客户之前，也可以用过滤器处理响应。

1.过滤器的工作原理

过滤器的工作原理如图 7-2 所示。

过滤器也在容器中，当一个 Java 类实现 javax.servlet.Filter 接口，成为过滤器后，则可以访问 ServletContext。

在 web.xml 中声明定义多个过滤器及其运行的顺序。

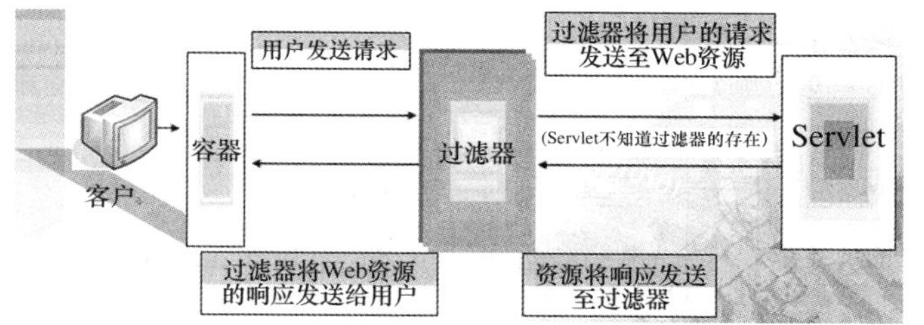

图 7-2　过滤器的工作原理

2.过滤器生命周期

(1)首先要有一个 init()

当容器实例化一个过滤器时,在 init()方法中完成过滤器所有初始化任务。

(2)真正的工作在 doFilter()中完成

容器对当前请求应用过滤器时,会调用 doFilter()方法完成过滤功能。该方法有三个参数:doFilter(ServletRequest,ServletResponse,FilterChain)。

(3)最后调用 destroy()

容器删除一个过滤器实例时,调用 destroy()方法进行回收清理工作。

3.声明和确定过滤器顺序

(1)在 web.xml 中声明过滤器

```
<filter>
  <filter-name>  </filter-name>
  <filter-class>  </filter-class>
  <init-param>{可选}
    <param-name>  </param-name>
    <param-value>   </param-value>
  </init-param>
</filter>
```

(2)将过滤器映射到想要过滤的 Web 资源。声明对应 URL 模式的过滤器映射:

```
①<filter-mapping>
  <filter-name>   </filter-name>
  <url-pattern>   </url-pattern>
  </filter-mapping>
②<filter-mapping>
    <filter-name>  </filter-name>
    <servlet-name>  </servlet-name>
  </filter-mapping>
```

注意:先找到与 URL 模式匹配的所有过滤器,最后再找与 servlet-name 匹配的过滤器。

根据 Servlet 2.3 规范,Filter 是按照 web.xml 配置的 filter-mapping 先后顺序执行的。

4.请求过滤与响应过滤顺序

请求过滤与响应过滤顺序如图 7-3 所示。

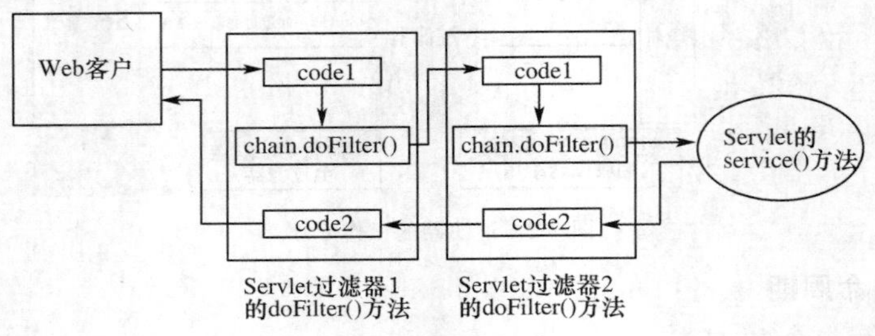

图 7-3　请求过滤与响应过滤顺序

7.4 实验内容与步骤

1.实验要求

要求使用 Servlet 过滤器编写程序,完成如下功能:

设计 3 个过滤器分别代表加、减、乘;提供输入框,用户输入数据,单击【submit】按钮,该数据经过上述过滤器,得到字符串表达式并输出表达式结果(加、减、乘过滤器初始化参数分别为 2、6、5)。

(1)由 MyServlet 生成用户输入界面,接收用户输入的数据,输入后单击【submit】按钮提交,如图 7-4 所示。

图 7-4　用户输入界面

(2)ResultServlet 接收由过滤器处理的字符串表达式,计算并输出结果,如图 7-5 所示。

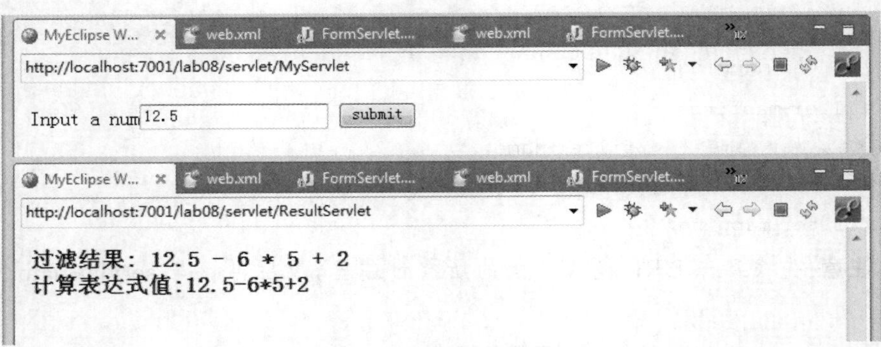

图 7-5　程序计算结果

2.项目结构如图 7-6 所示

图 7-6 项目结构图

3.定义过滤器

设计过滤器 AddFilter、SubFilter 和 MulitiFilter，实现 Filter 接口，实现 init()、doFilter()、destroy()接口方法。

```
package myFilter;
import java.io.IOException;
import java.util.Vector;
import javax.servlet.*;
public class AddFilter implements Filter{
    private FilterConfig filterConfig;
    public void init(FilterConfig filterConfig) throws ServletException{
        this.filterConfig=filterConfig;
    }
    public void destroy(){
    }
    public void doFilter(ServletRequest request,ServletResponse response,
        FilterChain filterChain) throws IOException,ServletException{
        Vector vc=new Vector();
        if(((request.getParameter("param")==null)||
        ((request.getParameter("param").length()==0)))){
            filterConfig.getServletContext().setAttribute("result",vc);
        }
        else{
            Vector mvc=(Vector)filterConfig.getServletContext().getAttribute("result");
            if(mvc.size()==0){
                mvc.add(request.getParameter("param"));
            }
```

```
            mvc.add("+");
            mvc.add(filterConfig.getInitParameter("add"));
            filterConfig.getServletContext().setAttribute("result",mvc);
        }
        try{
            filterChain.doFilter(request,response);
        }catch(ServletException sx){
            filterConfig.getServletContext().log(sx.getMessage());
        }catch(IOException iox){
            filterConfig.getServletContext().log(iox.getMessage());
        }
    }
}
```

SubFilter 过滤器类

```
package myfilter;
importjava.io.IOException;
importjava.util.Vector;
importjavax.servlet.Filter;
importjavax.servlet.FilterChain;
importjavax.servlet.FilterConfig;
importjavax.servlet.ServletException;
importjavax.servlet.ServletRequest;
importjavax.servlet.ServletResponse;
public class SubFilter implements Filter{
    privateFilterConfigfilterConfig;

    public void init(FilterConfig fc) throws ServletException {
        this.filterConfig = fc;

    }
    public void doFilter(ServletRequest request, ServletResponse response,
            FilterChainfilterChain) throws IOException, ServletException {

        Vector vc= new Vector();
        if(((request.getParameter("param"))= = null) ||((request.getParameter
("param")).length()= = 0)){
            filterConfig.getServletContext().setAttribute("result", vc);
        }
        else{
            Vector mvc = (Vector)filterConfig.getServletContext().getAttribute("
result");
            if(mvc.size()= = 0){
```

```
                mvc.add(request.getParameter("param"));
            }
            mvc.add("- ");
            mvc.add(filterConfig.getInitParameter("sub"));//12.5- 6
            filterConfig.getServletContext().setAttribute("result", mvc);
        }
        try{
            filterChain.doFilter(request, response);
        }catch(ServletExceptionsx){
            filterConfig.getServletContext().log(sx.getMessage());
        }catch(IOExceptioniox){
            filterConfig.getServletContext().log(iox.getMessage());
        }
    }

    public void destroy() {


    }
}
package myFilter;
import java.io.IOException;
import java.util.Vector;
import javax.servlet.* ;
public class MultiFilter implements Filter{
    private FilterConfig filterConfig;
    public void init(FilterConfig filterConfig) throws ServletException{
        this.filterConfig=filterConfig;
    }
    public void destroy(){
    }
    public void doFilter(ServletRequest request,ServletResponse response,
            FilterChain filterChain) throws IOException,ServletException{
        Vector vc=new Vector();
        if(((request.getParameter("param")==null)||
        ((request.getParameter("param").length()==0)))){
            filterConfig.getServletContext().setAttribute("result",vc);
        }
        else{
            Vector mvc=(Vector)filterConfig.getServletContext().getAttribute("result");
            if(mvc.size()==0){
                mvc.add(request.getParameter("param"));
```

```
            }
            mvc.add("*");
            mvc.add(filterConfig.getInitParameter("multi"));
            filterConfig.getServletContext().setAttribute("result",mvc);
        }
        try{
            filterChain.doFilter(request,response);
        }catch(ServletException sx){
            filterConfig.getServletContext().log(sx.getMessage());
        }catch(IOException iox){
            filterConfig.getServletContext().log(iox.getMessage());
        }
    }
}
```

4.定义 Servlet

设计 MyServlet,继承自 HttpServlet。

```
package com.myServlet;
import java.io.*;
import javax.servlet.*;
import javax.servlet.http.*;
public class MyServlet extends HttpServlet{
    private static final String CONTENT_TYPE="test/html;charset=GBK";
    public void init() throws ServletException{
    }
    public void doGet(HttpServletRequest request,HttpServletResponse response) throws
        ServletException,IOException{
        response.setContentType(CONTENT_TYPE);
        PrintWriter out=response.getWriter();
        out.println("<html>");
        out.println("<head><title>myServlet</title></head>");
        out.println("<body bgcolor=\"# ffffff\"");
        out.println("<form method=post action='/week8/servlet/ResultServlet'>");
        out.println("Input a num <input type=text name='param'/>");
        out.println("<input type=submit value=submit />");
        out.println("</form>");
        out.println("</body>");
        out.println("</html>");
        out.close();
    }
    public void doPost(HttpServletRequest request,HttpServletResponse response) throws
```

```
    ServletException,IOException{
        doGet(request,response);
    }
}
```

5.定义 ResultServlet,显示表达式和表达式运算结果

```
package myservlet;
import java.io.IOException;
import java.io.PrintWriter;
import java.util.Vector;
import javax.servlet.ServletException;
import javax.servlet.http.HttpServlet;
import javax.servlet.http.HttpServletRequest;
import javax.servlet.http.HttpServletResponse;
import bsh.Interpreter;
public class ResultServlet extends HttpServlet{
    public void doGet(HttpServletRequest request,HttpServletResponse response)
            throws ServletException,IOException{
        response.setContentType("text/html;charset=GBK");
        PrintWriter out=response.getWriter();
        out.println("<!DOCTYPE HTML PUBLIC \"-//W3C//DTD HTML 4.01 Transitional//EN\">");
        out.println("<HTML>");
        out.println("<HEAD><TITLE>A Servlet</TITLE></HEAD>");
        out.println("<BODY>");
        out.println("<h3>过滤结果: ");
        Vector vc=new Vector();
        vc=(Vector)getServletContext().getAttribute("result");
        String str=vc.elementAt(0).toString();
        out.println(vc.get(0));
        for(int i=1;i<vc.size();i++){
            str=str+ vc.elementAt(i);
            out.println(vc.get(i));
        }
        out.println("<br>"+"计算表达式值:"+str);
        Interpreter bsh=new Interpreter();
        try{
            bsh.eval("Result="+str.toString());
            out.println("="+bsh.get("Result"));
            vc.removeAllElements();
        }catch(Exception e){
```

```
        }
        out.println("</BODY>");
        out.println("</HTML>");
        out.flush();
        out.close();
    }
    public void doPost(HttpServletRequest request,HttpServletResponse response)
            throws ServletException,IOException{
        doGet(request,response);
    }
}
```

6.web.xml

定义 Filter、Filter 的映射及顺序,定义 Servlet 及映射。

```
<?xml version="1.0" encoding="UTF-8"? >
<web-app version="2.5"
    xmlns="http://java.sun.com/xml/ns/javaee"
    xmlns:xsi="http://www.w3.org/2001/XMLSchema-instance"
    xsi:schemaLocation="http://java.sun.com/xml/ns/javaee
    http://java.sun.com/xml/ns/javaee/web-app_2_5.xsd">
  <filter>
      <filter-name>MultiFilter</filter-name>
      <filter-class>com.myFilter.MultiFilter</filter-class>
      <init-param>
          <param-name>multi</param-name>
          <param-value>5</param-value>
      </init-param>
  </filter>
  <filter>
      <filter-name>AddFilter</filter-name>
      <filter-class>com.myFilter.AddFilter</filter-class>
      <init-param>
          <param-name>add</param-name>
          <param-value>2</param-value>
      </init-param>
  </filter>
  <filter>
      <filter-name>SubFilterFilter</filter-name>
      <filter-class>com.myFilter.SubFilter</filter-class>
      <init-param>
```

```
            <param-name>sub</param-name>
            <param-value>6</param-value>
        </init-param>
    </filter>
    <filter-mapping>
        <filter-name>AddFilter</filter-name>
        <servlet-name>ResultServlet</servlet-name>
    </filter-mapping>
    <filter-mapping>
        <filter-name>SubFilter</filter-name>
        <servlet-name>/*</servlet-name>
    </filter-mapping>
    <filter-mapping>
        <filter-name>MultiFilter</filter-name>
        <servlet-name>/*</servlet-name>
    </filter-mapping>
    <servlet>
      <description>This is the description of my J2EE component</description>
      <display-name>This is the display name of my J2EE component</display-name>
      <servlet-name>ResultServlet</servlet-name>
      <servlet-class>com.myServlet.ResultServlet</servlet-class>
    </servlet>
    <servlet>
      <description>This is the description of my J2EE component</description>
      <display-name>This is the display name of my J2EE component</display-name>
      <servlet-name>MyServlet</servlet-name>
      <servlet-class>com.myServlet.MyServlet</servlet-class>
    </servlet>
    <servlet-mapping>
      <servlet-name>ResultServlet</servlet-name>
      <url-pattern>/servlet/ResultServlet</url-pattern>
    </servlet-mapping>
    <servlet-mapping>
      <servlet-name>MyServlet</servlet-name>
      <url-pattern>/servlet/MyServlet</url-pattern>
    </servlet-mapping>
    <welcome-file-list>
      <welcome-file>index.jsp</welcome-file>
    </welcome-file-list>
</web-app>
```

7.部署项目，启动 WebLogic 服务器，运行 Servlet 程序

程序运行结果如图 7-7 所示。

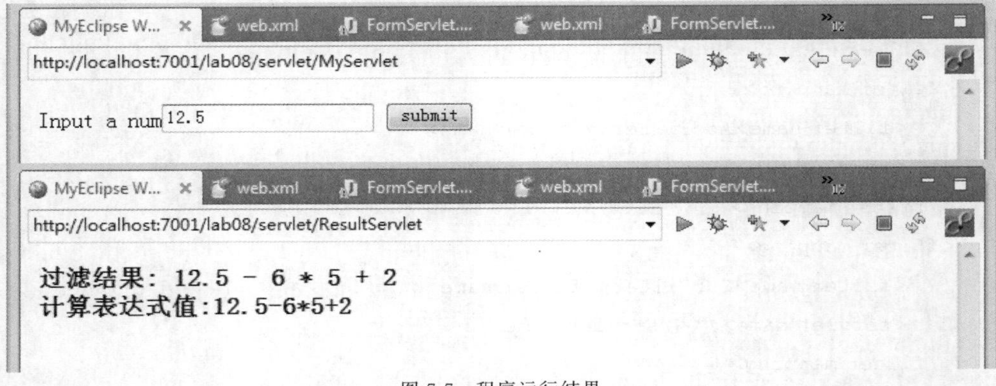

图 7-7 程序运行结果

7.5 实验总结

本实验从 Servlet 的层次结构入手，了解 Filter 生命周期及其接口包含的各种方法，如 init()、doFilter()和 destroy()，了解 Servlet 类层次结构，以更好地开发 Servlet 程序。对于客户端的某些请求，可以通过过滤器对其进行过滤，服务器端的响应结果也可以通过过滤器传给客户端。另外讲解了过滤器的创建、部署和运行的操作流程。

7.6 课后思考题

1.如何创建线程安全的 Servlet 程序？

2.简述 Servlet 编程中过滤器的作用。

3.简述过滤器包含的类和接口。

4.简述创建一个过滤器程序的实现步骤。

5.如何在 web.xml 文件中配置 Servlet 过滤器？

实验 8 Servlet 事件监听

8.1 实验目的

1.掌握定义一个 Servlet 并在 Web 应用程序内生成事件的方法
2.理解上下文、会话及请求不同监听接口的功能
3.掌握描述 Servlet 过滤器的方法
4.掌握利用监听器编写 Servlet 应用的步骤和方法

8.2 实验环境

1.MyEclipse 插件平台
2.WebLogic(或 Tomcat)容器

8.3 实验知识背景

8.3.1 Servlet 事件

在 Web 应用程序中,希望知道是否创建或撤销了一个上下文,或者上下文中是否增加、删除或替换了一个属性;想跟踪活动的会话,跟踪会话什么时候增加、删除或替换一个会话属性,跟踪某个对象什么时候绑定到一个会话中或从会话中删除;在每次请求到来时都希望监听到该请求,以便建立日志记录,并跟踪什么时候增加、删除或替换一个请求属性等,这些都依赖于 Servlet 事件监听和事件响应机制。针对上述需求,Servlet 有三类主要事件与之相对应,分别为 Servlet 上下文事件、会话事件和请求事件,可以针对上述事件编写事件监听器,对事件做出相应处理。

图 8-1 所示是 Core java 事件委托模型，Servlet 的事件处理与其相比，既有相同之处，又有不同之处。

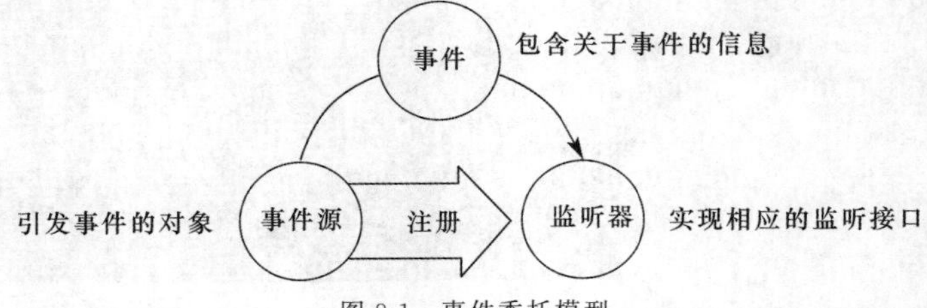

图 8-1　事件委托模型

1.事件源：主要讨论 ServletContext、Session、ServletRequest 三类对象及它们的属性。

2.事件：上述对象的产生、销毁及属性变更，以及 Session 的绑定等。

3.监听器：实现相应监听接口。Servlet API 定义了 8 个监听器接口，具体分类如下：

(1) ServletContextListener：监听 Servlet 上下文的产生与销毁。

(2) HttpSessionListener：监听 Session 的产生与销毁。

(3) ServletRequestListener：监听 Request 的产生与销毁。

(4) ServletContextAttributeListener：监听 Servlet 上下文属性的增加、修改与删除。

(5) HttpSessionAttributeListener：监听 Session 属性的增加、修改与删除。

(6) ServletRequestAttributeListener：监听 Request 属性的增加、修改与删除。

(7) HttpSessionActivationListener：监听 Session 对象的钝化与激活。

(8) HttpSessionBindingListener：监听一个对象绑定到 Session 或从 Session 中删除。

8.3.2　对 Servlet 上下文进行监听

在 Web 应用中可以部署监听程序，使其能够监听 ServletContext 上下文的信息（ServletContext 的创建和删除，ServletContext 属性的增加、删除和修改）。监听程序需要实现 ServletContextListener 和 ServletContextAttributeListener 接口，事件类型分别为 ServletContextEvent 和 ServletContextAttributeEvent。

ServletContextListener 接口：

```
public void contextInitialized(ServletContextEvent)
public void contextDestroyed(ServletContextEvent)
```

ServletContextAttributeListener 接口：

```
public void attributeAdded(ServletContextAttributeEvent)
public void attributeRemoved(ServletContextAttributeEvent)
public void attributeReplaced(ServletContextAttributeEvent)
```

注册监听器：可以如下所示在 web.xml 中指定，或无需指定，由容器自动检测。

在 web.xml 中部署监听器：

```
<listener>
    <listener-class>
        ……
```

```
    </listener-class>
</listener>
```

注意：不要放在<servlet>元素里。

8.3.3 监听 Http 会话

在 Web 应用中可以部署监听程序，监听 Http 会话相关的信息（会话活动情况，会话中属性设置情况，对象绑定到会话情况）。

监听程序需要实现以下接口：

1.HttpSessionListener 接口：监听 Http 会话的创建及销毁。

```
public void sessionCreated(HttpSessionEvent)
public void sessionDestroyed(HttpSessionEvent)
```

2.HttpSessionAttributeListener 接口：监听 Http 会话中属性的设置。

```
public void attributeAdded(HttpSessionBindingEvent)
public void attributeRemoved(HttpSessionBindingEvent)
public void attributeReplaced(HttpSessionBindingEvent)
```

注意：HttpSessionBindingEvent 中的方法 getValue()返回属性改变之前的值。

3.HttpSessionBindingListener 接口：监听对象本身绑定到一个会话或从会话中删除。

```
public void valueBound(HttpSessionBindingEvent)
public void valueUnbound(HttpSessionBindingEvent)
```

注意：

(1)HttpSessionAttributeListener：监听会话本身是否有属性增加、删除或修改。

(2)HttpSessionBindingListener：监听对象本身是否添加、删除于会话中。

HttpSessionBindingListener 必须实例化后放入某一个 Session 中，才可以进行监听。

有关监听 Session 绑定的事件监听器有以下几点需要注意：

保存到 Session 域中的对象有多种状态，包括：

绑定：session.setAtrribute("属性名",对象)；

解除绑定：session.removeAtrribute("属性名")，或 Session 过期或失效；

钝化：随 Session 对象持久化到一个存储设备；

活化：随 Session 对象从存储设备中恢复。

如果对象实现了 HttpSessionBindingListener、HttpSessionActivateListener 接口，那么，该对象处于绑定/松绑、钝化/活化时，就能自己感知到状态的变化，调用相应的方法。

8.3.4 对请求监听

在 Servlet 2.4 规范中，增加了一项技术，就是可以监听客户端的请求。监听程序需要实现以下接口：

1.ServletRequestListener(可以监听每次请求到来)

```
public void requestInitialized(ServletRequestEvent)
public void requestDestroyed(ServletRequestEvent)
```

2.ServletRequestAttributeListener(可以跟踪请求属性的增加、删除、修改)

```
public void attributeAdded(ServletRequestAttributeEvent)
public void attributeRemoved(ServletRequestAttributeEvent)
public void attributeReplaced(ServletRequestAttributeEvent)
```

8.4 实验内容与步骤

8.4.1 设计 Servlet 上下文事件监听器

设计 Servlet 上下文事件监听器，监听创建或撤销一个上下文和上下文中是否增加、删除或替换了一个属性事件。利用上下文监听器，将 web.xml 中的上下文初始参数构造一个 Dog 对象，将其存放于 ServletContext 的属性"dogname"中；编写 Servlet 输出该上下文的属性，当单击【删除上下文属性】按钮时，该属性被删除，同时上下文监听器监听出其属性被删除。

1.项目结构如图 8-2 所示

图 8-2 项目结构图

2.配置 web.xml

在 web.xml 中分别配置 DelServletContextAttr、DispServletContext servlet 及 servlet-mapping，设置上下文初始参数 breed，注册 listener。

```
<context-param>
    <param-name>breed</param-name>
    <param-value>Great Dane</param-value>
  </context-param>
```

```
<listener>
  <listener-class>com.myListener.MyservletContextListener</listener-class>
</listener>
<servlet>
  <description>This is the description of my J2EE component</description>
  <display-name>This is the display name of my J2EE component</display-name>
  <servlet-name>DelServletContextAttr</servlet-name>
  <servlet-class>com.myServlet.DelServletContextAttr</servlet-class>
</servlet>
<servlet>
  <description>This is the description of my J2EE component</description>
  <display-name>This is the display name of my J2EE component</display-name>
  <servlet-name>DispServletContext</servlet-name>
  <servlet-class>com.myServlet.DispServletContext</servlet-class>
</servlet>
<servlet-mapping>
  <servlet-name>DelServletContextAttr</servlet-name>
  <url-pattern>/servlet/DelServletContextAttr</url-pattern>
</servlet-mapping>
<servlet-mapping>
  <servlet-name>DispServletContext</servlet-name>
  <url-pattern>/servlet/DispServletContext</url-pattern>
</servlet-mapping>
```

3.定义上下文监听器

设计监听器 MyServletContextListener，实现 ServletContextAttributeListener、ServletContextListener 接口，应用对应的接口方法 valueBound(HttpSessionBindingEvent)、valueUnbound(HttpSessionBindingEvent)监听上下文及上下文属性事件。

```
package com.myListener;
import javax.servlet.*;
import com.myPackage.Dog;
public class MyServletContextListener implements
    ServletContextAttributeListener,ServletContextListener{
    public void attributeAdded(ServletContextAttributeEvent scab){
        System.out.println("WEB上下文中添加属性:"+ scab.getName()+ ":"+ scab.getValue());
    }
    public void attributeRemoved(ServletContextAttributeEvent scab){
        System.out.println("WEB上下文中删除属性:"+ scab.getName()+ ":"+ scab.getValue());
    }
    public void attributeReplaced(ServletContextAttributeEvent scab){
        // TODO Auto-generated method stub
```

```
    }
    public void contextDestroyed(ServletContextEvent arg0){
        // TODO Auto-generated method stub
    }
    public void contextInitialized(ServletContextEvent sce){
        ServletContext sc=sce.getServletContext();
        String str=sc.getInitParameter("breed");
        Dog d=new Dog(str);
        sc.setAttribute("dogname",d);
    }
}
```

4.定义 Dog 类

封装 JavaBean Dog 类，在 web.xml 中设置上下文初始参数：breed＝Great Dane。

```
package com.myPackage;
public class Dog{
    private String breed;
    public Dog(String s){
        breed=s;
    }
    public String getBreed(){
        return breed;
    }
}
```


5.定义 Servlet

设计 servlet DelServletContextAttr、DispServletContext，分别添加和删除上下文属性，引发上下文事件。

```
package com.myServlet;
import javax.servlet.http.HttpServlet;
import javax.servlet.http.*;
import javax.servlet.*;
import java.io.*;
public class DelServletContextAttr extends HttpServlet{
    private static final String CONTEXT_TYPE="text/html;charset=GBK";
    public void init() throws ServletException{
    }
    public void doGet(HttpServletRequest request,HttpServletResponse response) throws
        IOException,ServletException{
        response.setContentType(CONTEXT_TYPE);
        PrintWriter out=response.getWriter();
        out.println("<html>");
```

```
        out.println("<head><title>deleteServlet</title></head>");
        out.println("<body bgcolor=\"# ffffff\">");
        getServletContext().removeAttribute("dogname");
        out.println("<h3>上下文属性删除成功!");
        out.println("</body>");
        out.println("</html>");
        out.close();
    }
    public void getPost(HttpServletRequest request,HttpServletResponse response)throws
        IOException,ServletException{
        doGet(request,response);
    }
    public void destroy(){
    }
}
package com.myServlet;
import javax.servlet.http.HttpServlet;
import javax.servlet.*;
import javax.servlet.http.*;
import java.io.*;
import com.myPackage.*;
public class DispServletContext extends HttpServlet{
    private static final String CONTEXT_TYPE="text/html;charset=GBK";
    public void init() throws ServletException{
    }
    public void doGet(HttpServletRequest request,HttpServletResponse response) throws
        IOException,ServletException{
        response.setContentType(CONTEXT_TYPE);
        PrintWriter out=response.getWriter();
        out.println("</html>");
        out.println("<head><title>myServlet</title></head>");
        out.println("<body bgcolor=\"# ffffff\"");
        out.println("<h3>测试上下文监听器<br>");
        Dog d=(Dog)getServletContext().getAttribute("dogname");
        out.println("dog's breed is:"+ d.getBreed());
        out.println("<form method=post action='/week9-1/servlet/DelServletContextAttr'>");
        out.println("<input type='submit' valuse='删除上下文属性'/>");
        out.println("</form>");
        out.println("</body>");
        out.println("</html>");
        out.close();
```

```
    }
    public void doPost(HttpServletRequest request,HttpServletResponse response) throws
        IOException,ServletException{
        doGet(request,response);
    }
    public void destroy(){
    }
}
```

6.部署项目,启动 WebLogic 服务器并运行 Servlet 程序

程序运行结果如图 8-3 所示。

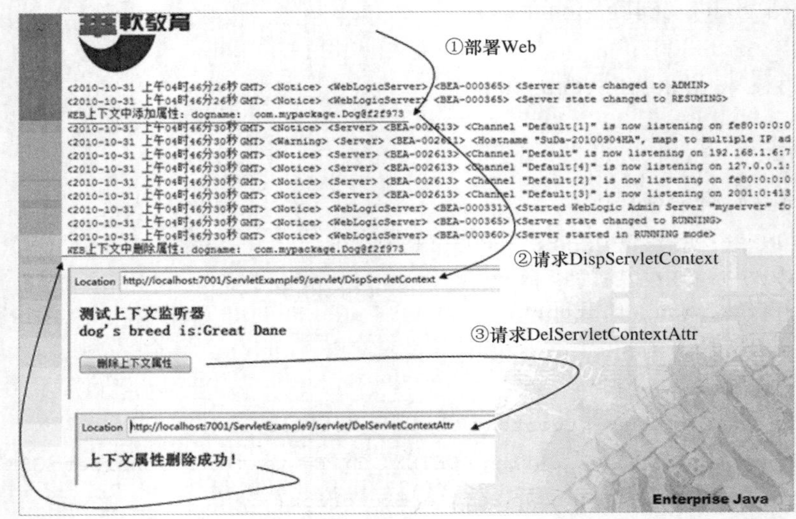

图 8-3　程序运行结果

8.4.2　设计监听器,允许属性跟踪事件

设计监听器,允许属性跟踪可能对其本身很重要的事件,如对象增加到会话或从会话中删除等。实现学生对象属性和数据库表记录实时更新功能,需求如图 8-4 所示。

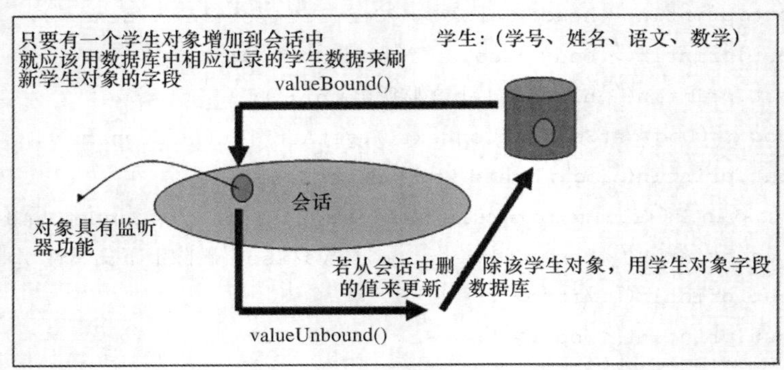

图 8-4　用户需求示意图

1.项目结构,如图 8-5 所示

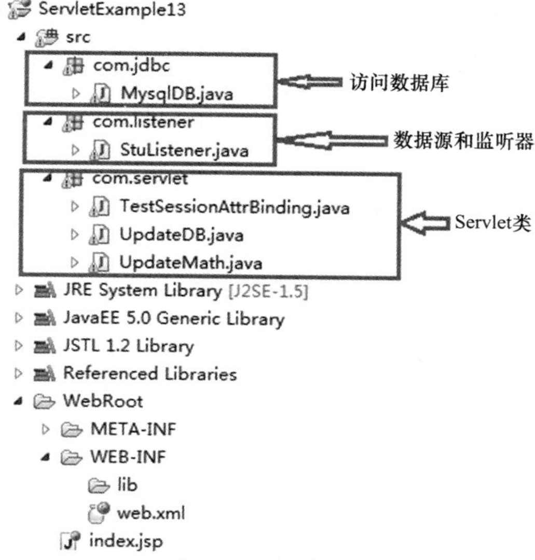

图 8-5 项目结构图

2.设计监听器 StuListener

设计监听器 StuListener 实现 HttpSessionBindingListener 接口,应用如下接口方法 valueBound(HttpSessionBindingEvent)、valueUnbound(HttpSessionBindingEvent)监听对象本身绑定到一个会话或从会话中删除。

StuListener 类

```
package com.listener;
import javax.servlet.* ;
import java.util.* ;
import javax.servlet.http.HttpSessionBindingEvent;
import javax.servlet.http.HttpSessionBindingListener;
import com.jdbc.MysqlDB;
public class StuListener implements HttpSessionBindingListener{
    private int sno;
    private String sname;
    private int chinese;
    private int math;
    public StuListener(int i1){
        setSno(i1);}
    public StuListener(int i1,String s,int i2,int i3){
        setSno(i1);
        setSname(s);
        setChinese(i2);
        setMath(i3);}
    public String _toString(){ return sno+ ""+ sname+ ""+ chinese+ ""+ math; };
```

```
    public void setSno(int sno){ this.sno= sno; };
    public int getSno(){ return sno; };
    public void setSname(String sname){ this.sname= sname; };
    public String getSname(){ return sname; };
    public void setChinese(int chinese){ this.chinese= chinese; };
    public int getChinese(){ return chinese; };
    public void setMath(int math){ this.math= math; };
    public int getMath(){ return math; };
  public void valueBound(HttpSessionBindingEvent event){
    System.out.println("对象绑于会话中....");
    MysqlDB mdb =  new MysqlDB(sno);
    StuListener st =  mdb.selectDBByid();
    sname =  st.getSname();
    chinese =  st.getChinese();
    math =  st.getMath(); }
  public void valueUnbound(HttpSessionBindingEvent arg0){
    System.out.println("从会话中删除对象....");
    MysqlDB mdb =  new MysqlDB(sno);
    System.out.println(mdb.updateDB(math));} }
```

3.定义 servlet

设计 servlet TestSessionAttrBinding、UpdateDB、UpdateMath，使用到对象邦定到会话中的事件。（注意：HttpSessionBindingListener 必须实例化后放入某一个 session 中，才可以进行监听，不需要在 web.xml 中去配置）。

TestSessionAttrBinding 类

```
package com.servlet;
import java.io.IOException;
import java.io.PrintWriter;
import java.util.* ;
import javax.servlet.ServletException;
import javax.servlet.http.* ;
import com.jdbc.MysqlDB;
import com.listener.StuListener;
public class TestSessionAttrBinding extends HttpServlet{
  private static final String CONTEXT_TYPE =  "text/html;charset= GBK";
  public void init() throws ServletException{ }
  public void doGet(HttpServletRequest request,HttpServletResponse response)
  throws ServletException,IOException{
    response.setContentType(CONTEXT_TYPE);
    PrintWriter out =  response.getWriter();
    MysqlDB mdb =  new MysqlDB();
```

```
            out.println("< html> ");
            out.println("< head> < title> TestSessionAttributeBinding< /title>
< /head> ");
            out.println("< body bgcolor= \"# ffffff\"> ");
            out.println("< h3> 测试对象绑定会话监听");
            out. println (" < form name = form1 method = post action ='/
ServletExample13/servlet/UpdateDB'> ");
            out.println("< 数据库记录信息> ");
            out.println("< select name= 'select' size= 10> ");
            Vector vc = mdb.selectDB();
            for(int i= 0;i< vc.size();i+ + ){
                out.println("< option value= "+ ((StuListener)vc.get(i)).getSno()+ "> ");
                out.println(((StuListener)vc.get(i))._toString());
                out.println("< /option> ");
                }
            out.println("< /select> ");
            out.println("< hr> ");
            out.println("< input type= submit value= '选择'/> ");
            out.println("< /form> ");
            out.println("< /html> ");
            out.close();
        }
        public void doPost ( HttpServletRequest request, HttpServletResponse
response) throws
        ServletException,IOException{
            doGet(request,response);    }}
```

(C) UpdateDB类

```
    package com.servlet;
    import java.io.IOException;
    import java.io.PrintWriter;
    import javax.servlet.ServletException;
    import javax.servlet.http.HttpServlet;
    import javax.servlet.http.HttpServletRequest;
    import javax.servlet.http.HttpServletResponse;
    import javax.servlet.http.HttpSession;
    import com.listener.StuListener;
    public class UpdateDB extends HttpServlet{
        public void init() throws ServletException{}
        protected void doGet(HttpServletRequest req, HttpServletResponse resp)
                throws ServletException, IOException {
            resp.setContentType("text/html; charset= GBK");
            PrintWriter out = resp.getWriter();
```

```
        req.setCharacterEncoding("GBK");
        out.println("< html> ");
        out.println("< head> < title> TestSessionAttributeBinding< /title>
< /head> ");
        out.println("< body bgcolor= \"# ffffff\"> ");
        out.println("< h3> 修改选择记录:");
        StuListener stu =  new StuListener(Integer.parseInt(req.getParameter
("select")));
        HttpSession session= req.getSession();
        session.setAttribute("stu", stu);
        out.println("< form method= post action= '/lab9/servlet/UpdateMath'> ");
        out.println("学号"+ stu.getSno() + "姓名"+ stu.getSname() + "语文"+ stu.
getChinese() + "< br> ");
        out.println("数学");
        out.println("< input type= text name= 'math' value= "+ stu.getMath()
+"> ");
        out.println("< input type= submit value= '修改数学成绩'/> ");
        out.println("< /form> ");
        out.println("< /body> ");
        out.println("< /html> ");
        out.close();
    }
    protected void doPost(HttpServletRequest req, HttpServletResponse resp)
            throws ServletException, IOException {
        // TODO Auto-generated method stub
        doGet(req, resp);
    }
}
```

UpdateDB 类

```
package com.servlet;
import java.io.IOException;
import java.io.PrintWriter;
import java.util.* ;
import javax.servlet.ServletException;
import javax.servlet.http.* ;
import com.jdbc.MysqlDB;
import com.listener.StuListener;
public class UpdateDB extends HttpServlet{
    private static final String CONTEXT_TYPE = "text/html;charset= GBK";
    public void init() throws ServletException{ }
    public  void  doGet  ( HttpServletRequest  request,  HttpServletResponse
```

```
response)
        throws ServletException,IOException{
        response.setContentType(CONTEXT_TYPE);
        PrintWriter out = response.getWriter();
        out.println("< html> ");
        out.println("< head> < title> TestSessionAttributeBinding< /title> < /
head> ");
        out.println("< body bgcolor= \"# ffffff\"> ");
        out.println("< h3> 修改选择记录");
        StuListener stu = new StuListener(Integer.parseInt(request.getParameter
("select")));
        HttpSession session = request.getSession();
        session.setAttribute("stu", stu);
        out. println (" < form method = post action = '/ServletExample13/servlet/
UpdateMath'> ");
        out.println("学号:"+ stu.getSno() + " 姓名:"+ stu.getSname() + " 语文:"+ stu.
getChinese() + "< br> ");
        out.println("数学");
        out.println("< input type= text name= 'math' value= "+ stu.getMath() + "> ");
        out.println("< input type= submit value= '修改数学成绩'/> ");
        out.println("< /form> ");
        out. println (" < br > < form method = post action = '/ServletExample13/
servlet/TestSessionAttrBinding'> ");
        out.println("< input type= submit value= '返回首页'/> ");
        out.println("< /form> ");
        out.println("< /body> ");
        out.println("< /html> ");
        out.close(); }
        public void doPost ( HttpServletRequest request, HttpServletResponse
response) throws
        ServletException,IOException{
            doGet(request,response); }
    }
```

UpdateMath 类

```
package com.servlet;
import java.io.IOException;
import java.io.PrintWriter;
import java.util.* ;
import javax.servlet.ServletException;
import javax.servlet.http.HttpServlet;
import javax.servlet.http.HttpServletRequest;
```

```
import javax.servlet.http.HttpServletResponse;
import javax.servlet.http.HttpSession;
import com.listener.StuListener;
public class UpdateMath extends HttpServlet{
    private static final String CONTEXT_TYPE = "text/html;charset= GBK";
    public void init() throws ServletException{ }
    public void doGet(HttpServletRequest request,HttpServletResponse response)
            throws ServletException,IOException{
        response.setContentType(CONTEXT_TYPE);
        PrintWriter out = response.getWriter();
        out.println("< html> ");
        out.println("< head> < title> UpdateMath< /title> < /head> ");
        out.println("< body bgcolor= \"# ffffff\"> ");
        out.println("< h3> 修改数学成绩完成");
        HttpSession session = request.getSession();
        StuListener st= (StuListener)session.getAttribute("stu");
        if(request.getParameter("math")! = null)
        st.setMath(Integer.parseInt(request.getParameter("math")));
        session.removeAttribute("stu");
        out.println ("< br> < form method= post action= '/ServletExample13/
servlet/TestSessionAttrBinding'> ");
        out.println("< input type= submit value= '返回首页'/> ");
        out.println("< /form> ");
        out.println("< /body> ");
        out.println("< /html> ");
        out.close();
        }
    public void doPost (HttpServletRequest request, HttpServletResponse
response) throws
    ServletException,IOException{
        doGet(request,response);
    }
}
```

4.定义 Java Bean MysqlDB。

在 java Bean MysqlDB 中定义通过 JNDI 查找数据源、对数据库进行增加、删除、修改的操作的方法，首先需要在数据库服务器中创建相关数据库和关系表。然后在 weblogic console 中配置 JDBC 数据源，建立 JNDI 名称。

MysqlDB 类

```
package com.jdbc;
```

```
import java.sql.* ;
import java.util.* ;
import javax.naming.* ;
import javax.sql.DataSource;
import com.listener.StuListener;
public class MysqlDB {
    int id;
    Vector vc;
    public MysqlDB(){
        vc = new Vector();
    }
    public MysqlDB(int i){
        id = i;
        vc = new Vector();
    }
    //访问数据库
    public Vector selectDB(){
        final String JNDI_DATABASE_NAME = "AA";
            Properties h = new Properties();
            h.put(InitialContext.INITIAL_CONTEXT_FACTORY,
                "weblogic.jndi.WLInitialContextFactory");
            String hostURL = "t3://localhost:7001";

            h.put(InitialContext.PROVIDER_URL, hostURL);
            try{
                Context ic = new InitialContext(h);
                DataSource ds = (DataSource) ic.lookup(JNDI_DATABASE_NAME);
                System.out.println("connect ok");
                Connection conn = ds.getConnection();
                String sql_string = "select * from stdb";
                PreparedStatement stmt = conn.prepareStatement(sql_string);
                ResultSet rs = stmt.executeQuery();
                while(rs.next()){
                    vc.add(new StuListener(rs.getInt("sno"),
                    rs.getString("sname"),rs.getInt("chinese"),rs.getInt("math")));
                }
                rs.close();
                stmt.close();
                conn.close();
                return vc;
            }catch(Exception e){e.printStackTrace();}
```

```
        return null;

    }
    //通过学号获取记录
    public StuListener selectDBByid(){
        final String JNDI_DATABASE_NAME = "AA";
            Properties h = new Properties();
            h.put(InitialContext.INITIAL_CONTEXT_FACTORY,
                  "weblogic.jndi.WLInitialContextFactory");
            String hostURL = "t3://localhost:7001";

            h.put(InitialContext.PROVIDER_URL, hostURL);
            StuListener st = null;
            try{
                Context ic = new InitialContext(h);
                DataSource ds = (DataSource) ic.lookup(JNDI_DATABASE_NAME);
                System.out.println("connect ok");
                Connection conn = ds.getConnection();
                String sql_string = "select * from stdb where sno= ?";
                PreparedStatement stmt = conn.prepareStatement(sql_string);
                stmt.setInt(1, id);
                ResultSet rs = stmt.executeQuery();
                while(rs.next()){
                    st = new StuListener(rs.getInt("sno"),
                    rs.getString("sname"),rs.getInt("chinese"),rs.getInt("math"));
                }
                rs.close();
                stmt.close();
                conn.close();
                return st;
            }catch(Exception e){e.printStackTrace();}
            return null;}
    public String updateDB(int test){
        final String JNDI_DATABASE_NAME = "AA";
            Properties h = new Properties();
            h.put(InitialContext.INITIAL_CONTEXT_FACTORY,
            "weblogic.jndi.WLInitialContextFactory");
            String hostURL = "t3://localhost:7001";
            h.put(InitialContext.PROVIDER_URL, hostURL);
            StuListener stb = selectDBByid();
            if(stb.getMath()! = test){
```

```
            try{Context ic = new InitialContext(h);
                DataSource ds = (DataSource) ic.lookup(JNDI_DATABASE_NAME);
                System.out.println("connect ok");
                Connection conn = ds.getConnection();
                String sql_string = "update stdb set math= ? where sno= ?";
                PreparedStatement stmt = conn.prepareStatement(sql_string);
                stmt.setInt(1, test);
                stmt.setInt(2, id);
                int row = stmt.executeUpdate();
                if(row > 0){
                    System.out.println("修改完成");
                    stmt.close();
                    conn.close();
                    return "修改完成";
                }else{
                    stmt.close();
                    conn.close();
                    return "修改不完成";}
            }catch(Exception e){
                e.printStackTrace();
            }
            return null;
        }else
            return "成绩没改变,不用修改!"; }
}
```

5.配置web.xml文件。

在web.xml中分别为配置TestSessionAttrBinding、UpdateDB、UpdateMath servlet及servlet-mapping。

```
<?xml version="1.0" encoding="UTF-8"?>
<web-app    xmlns:xsi="http://www.w3.org/2001/XMLSchema-instance" xmlns="http://java.sun.com/xml/ns/javaee" xsi:schemaLocation="http://java.sun.com/xml/ns/javaee http://java.sun.com/xml/ns/javaee/web-app_2_5.xsd" id="WebApp_ID" version="2.5">

  <servlet>
    <servlet-name>TestSessionAttrBinding</servlet-name>
    <servlet-class>com.servlet.TestSessionAttrBinding</servlet-class>
  </servlet>
  <servlet>
    <servlet-name>UpdateDB</servlet-name>
```

```
    < servlet-class> com.servlet.UpdateDB< /servlet-class>
  < /servlet>
  < servlet>
    < servlet-name> UpdateMath< /servlet-name>
    < servlet-class> com.servlet.UpdateMath< /servlet-class>
  < /servlet>
  < servlet-mapping>
    < servlet-name> TestSessionAttrBinding< /servlet-name>
    < url-pattern> /servlet/TestSessionAttrBinding< /url-pattern>
  < /servlet-mapping>
  < servlet-mapping>
    < servlet-name> UpdateDB< /servlet-name>
    < url-pattern> /servlet/UpdateDB< /url-pattern>
  < /servlet-mapping>
  < servlet-mapping>
    < servlet-name> UpdateMath< /servlet-name>
    < url-pattern> /servlet/UpdateMath< /url-pattern>
  < /servlet-mapping>
  < welcome-file-list>

    < welcome-file> index.jsp< /welcome-file>

  < /welcome-file-list>
< /web-app>
```

6.部署项目、启动 weblogic 服务器、运行 servlet 程序

程序运行结果如图 8-6 所示。

(1)选择修改学生

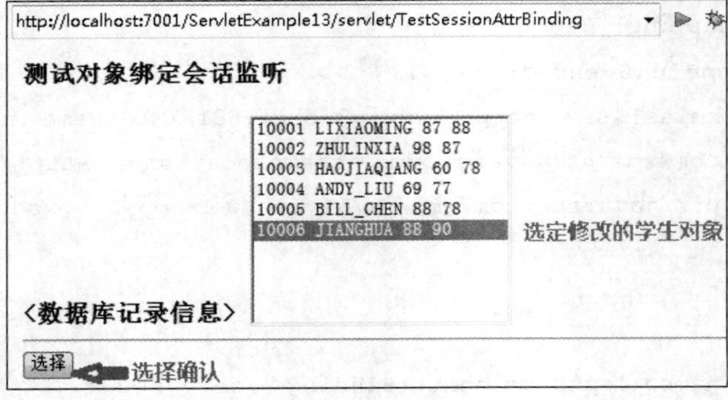

(2)修改学生数学成绩

http://localhost:7001/ServletExample13/servlet/UpdateDB

修改选择记录

学号:10006 姓名:JIANGHUA 语文:88

数学 90 修改数学成绩

返回首页

http://localhost:7001/ServletExample13/servlet/UpdateDB

修改选择记录

学号:10006 姓名:JIANGHUA 语文:88

数学 100 修改数学成绩

返回首页

(3)监听对象删除事件

http://localhost:7001/ServletExample13/servlet/UpdateMath

修改数学成绩完成

返回首页

```
connect ok
对象绑于会话中....
connect ok
从会话中删除对象....
connect ok
成绩没改变，不用修改!
<2018-12-17 下午05时26分39秒 CST>
从会话中删除对象....
connect ok
connect ok
修改完成
修改完成
<2018-12-17 下午05时28分49秒 CST>
```

(4)再次确认修改结果

图 8-6 程序运行结果

8.5 实验总结

本实验采用 Servlet 的特殊用法——监听器，监听器则是对于 Web 应用中的某些特殊事件，类似于 Java 图形界面编程中提供的事件处理机制，介绍了 servlet 事件的上下文、会话事件和请求事件对应的监听器的接口、接口提供的方法和方法的触发方式现对 Srvlet 三类主要事件编写事件监听器，对事件做出相应处理。

8.6 课后思考题

1.Servlet 的监听接口有哪些？

2.简述接口提供的方法在什么情况下会触发？

3.如何在 web.xml 文件中配置 Servlet？

4.如何在 web.xml 文件中配置 Servlet 监听器？

5.Servlet 的监听程序设计的基本步骤有哪些？

实验 9 JSP 技术基础知识

9.1 实验目的

1.了解 JSP 与 Servlet 的联系与区别

2.掌握 JSP 的基本组成：模板、page 指令、声明、表达式、脚本和注释

3.熟悉使用 MyEclipse 开发 JSP 的主要步骤

4.读懂 JSP 转换为 Servlet 程序的主要内容

9.2 实验环境

1.MyEclipse 插件平台

2.WebLogic(或 Tomcat)容器

9.3 实验知识背景

9.3.1 什么是 JSP

1.JSP 是 Java Server Pages 的缩写，是由 Sun 公司倡导的一种动态网页技术标准，可以建立安全、跨平台的动态网站。

2.JSP 是一种服务器端技术，能够将 Java 代码片断嵌入到 HTML 代码中。这些 Java 代码将生成动态内容，并嵌入到 HTML 页面中。

3.在 JSP 中既可以使用标准标签来嵌入代码，也可以使用定制标签和 JavaBean 来生成动态内容。

4.JSP 与 Servlet 密切相关，JSP 文件在用户第一次请求时都要编译成 Servlet，然后由该 Servlet 来处理用户请求。

5.JSP 文件的扩展名为.jsp。

6.使用 JSP 时，不需要单独配置每一个文件，JSP 容器(也就是 Servlet 容器)能够自动识别。

9.3.2 JSP 的执行过程

1.在 HTML 中嵌入 Java 脚本代码。

2.由应用服务器中的 JSP 引擎来编译和执行嵌入的 Java 脚本代码。

3.将生成的整个页面信息返回给客户端。

执行过程如图 9-1 所示。

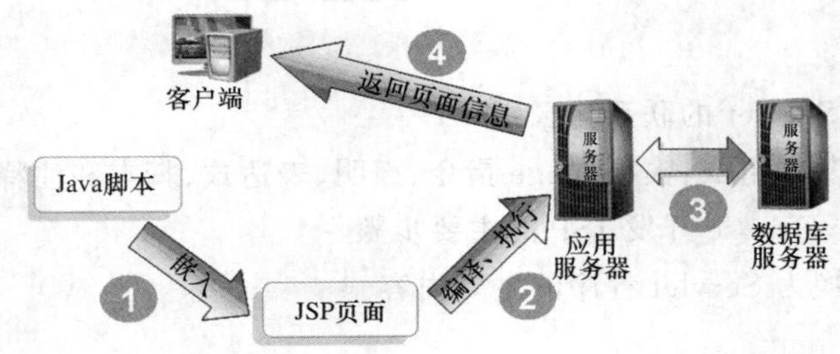

图 9-1 JSP 执行示意图

Web 容器处理 JSP 文件请求需要经过 3 个阶段，即翻译阶段、编译阶段和执行阶段。

客户端的请求(Request)通过 Web 服务器(Web Server)交给 JSP 引擎转换成 Java 的.class 文件，即 Servlet，之后 Servlet 引擎将其载入内存运行。运行结果(Response)以 HTML 或 XML 形式通过 Web 服务器返回给客户端，具体过程如图 9-2 所示。

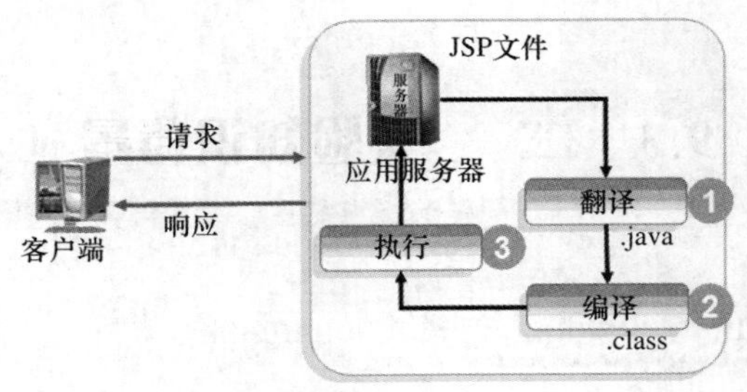

图 9-2 Web 容器处理 JSP 请求的过程

如图 9-3 所示，在第一次请求后，对于第二次请求或者后续的请求，Web 容器可以重用已经编译好的字节码文件，以后对该文件的访问就不需要再次编译了，这样能够提高后续的访问速度。但是，如果 JSP 文件发生变化，访问时则会重新编译。

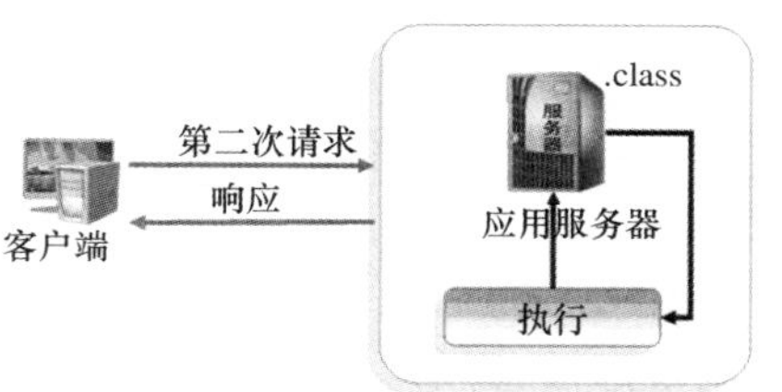

图 9-3 Web 容器处理第二次 JSP 请求的过程

9.3.3 JSP 的构成元素

构成元素一般是用 Java 编写的代码，可以嵌入到 JSP 页面中。下面介绍一下 JSP 的构成元素。JSP 的构成元素有：模板内容、指令、脚本元素、操作元素、注释和 EL 等。

1.模板内容

模板内容是指 JSP 页面中的静态 HTML 或 XML 内容。模板是一种 JSP 文件，包含参数化的内容，例如，template:get、template:put 和 template:insert。所有非 JSP 元素归为模板内容，包括所有静态内容。所有的模板内容都是可选的，可以很容易地在更多的网页中使用。

2.指令

指令在 JSP 翻译成 Servlet 期间提供整个 JSP 页面的相关信息，且不会产生任何输出信息到当前输出流中。

格式：<%@ 指令 属性名 1="…" 属性名 2="…" %>

说明：<与%、%与@、%与>之间不能出现空格

指令有三种：page、include 和 taglib，例如：

```
<%@page import="java.util.* "contentType="text/html;charset=GB2312"%>
```

3.脚本元素

脚本元素包括声明、表达式和 Scriptlet，用于将 Java 代码包含于 JSP 中。

(1)声明

格式 1：<% 变量表；%>　　声明某一方法的局部变量

格式 2：<%! 变量表；%>　　声明类的实例变量或类变量(带 Static 时)

例如：

```
<%int a=10;%>                声明局部变量
<%! int b=20;%>              声明类的实例变量
<%! Static int c=30;%>       声明类变量
```

(2)表达式

格式：<%=表达式 %>　　输出表达式的值

说明：表达式末尾不能有分号。

例如：<%=10+20%>

(3)Scriptlet (脚本段)

格式：<% Java 语句；>

例如：

```
<%for(int i=1;i<10;i++)
    out.println(i+"<br>");
%>
```

4.操作元素

操作元素为请求处理阶段提供信息(符合 XML 格式，即包含开始标签、属性或可选内容、结束标签，也可以是空标签和属性)。

例如：＜jsp:param name="名字" value="值"/＞

5.注释

格式：＜%--注释内容--%＞

JSP 将忽略注释内容。

6.EL

后续章节介绍。

9.3.4 JSP 的注释

JSP 中的注释有多种形式，既有 JSP 自带的注释规范，也有 HTML/XML 的注释规范，下面进行详细介绍。

1.HTML/XML 注释

此类注释经过响应输出流后不会改变，被包含在客户端 HTML 中，在浏览窗口中是不可见的，但可通过选择“查看源文件”命令来查看。

格式：＜!--comment[＜%=expression%＞]--＞

例如：＜!--这是一个典型的 JSP，它包含了 JSP 中常用的元素--＞

在客户端的源代码中产生和上面一样的代码：

```
<!--这是一个典型的 JSP,它包含了 JSP 中常用的元素-->
```

这种注释和 HTML 的注释类似，唯一不同之处就是可以在这种注释中使用表达式，例如：

```
<!--当前时间为:<% =(new java.util.Date()).toLocaleString()%>-->
```

在客户端的 HTML 源代码中显示如下：

```
<!--当前时间为:2017-1-02 17:30:24-->
```

2.隐藏注释

隐藏注释写在 JSP 文件中，但不发送到客户端，所以在客户端是不可见的。

格式：＜%--comment--%＞

例如：＜%-- 下面是使用表达式的例子--%＞

用隐藏注释标记的字符在 JSP 编译时将被忽略，JSP 编译器不会对“＜%--”和“--%＞”之间的语句进行编译，因此这些语句不会显示在客户端的浏览器中，也不会出现在客户端 HTML 源代码中。

3.Scriptlets 中的注释

由于包含的是 Java 代码，所以 Java 中的注释在 Scriptlets 中也适用，常用的 Java 注释中，“//”表示单行注释，“/ * * … * /”表示多行注释。

例如：

```
//color 表示颜色，通过它来动态控制颜色
```

也可以写成：

```
/**
    * color 表示颜色，通过它来动态控制颜色
*/
```

9.3.5 利用 MyEclipse 开发 JSP

在 MyEclipse 中，开发 JSP 与开发 Servlet 类似，但前者更简单，主要步骤如下：

1.创建 Web 工程。

2.新建 JSP 文件(【文件】|【新建】|【JSP】)，如图 9-4 所示。

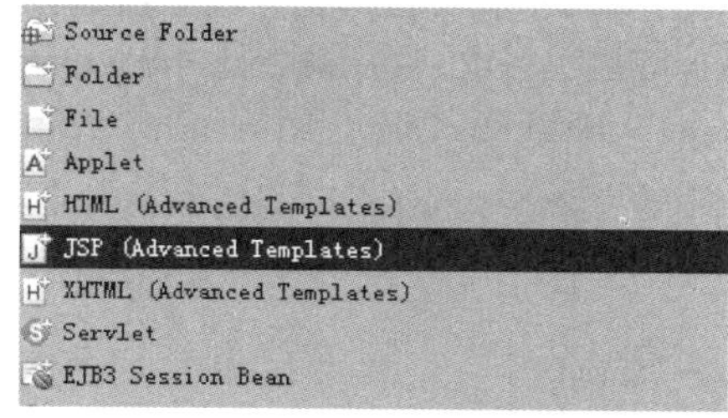

图 9-4 新建 JSP 文件

3.对 JSP 文件进行编码。

4.部署项目，Web 容器具有自动识别功能，不必对每一个 JSP 进行配置。

9.4 实验内容与步骤

1.按要求编写 JSP 程序，并回答相关问题。

该程序的功能是：显示局部变量和实例变量的不同，输出系统日期和九九表(下三角)。

(1)新建 Web 项目，在项目中新建 JSP 文件，命名为"myjsp.jsp"。

源代码如下：

```
<%@page contentType="text/html;charset=GB2312"%>
<%@page import="java.util.* "%>
<html>
    <head>
        <title>JSP 简单使用</title>
    </head>
    <body>
        <%
            int localvar=0;
```

```
%>
<%! int count=0;%>
<%
    localvar++;
    count++;
%>
局部变量 localvar=<%=localvar%><br>
实例变量 count=<%=count%><br>
<%
    //获取系统当前日期和时间值
    Calendar calendar=new GregorianCalendar();
    int year=calendar.get(Calendar.YEAR);
    int month=calendar.get(Calendar.MONTH)+1;
    int day=calendar.get(Calendar.DAY_OF_MONTH);
    int dayOfWeek=calendar.get(Calendar.DAY_OF_WEEK);
    String weekDay=null;
    switch (dayOfWeek){
    case 1:
        weekDay="星期日";
        break;
    case 2:
        weekDay="星期一";
        break;
    case 3:
        weekDay="星期二";
        break;
    case 4:
        weekDay="星期三";
        break;
    case 5:
        weekDay="星期四";
        break;
    case 6:
        weekDay="星期五";
        break;
    case 7:
        weekDay="星期六";
        break;
    }
%><p>今天是:<%=year%>年<%=month%>月<%=day%>日
<%=weekDay%>
<H4>
```

```
        打印九九表(下三角)
    </H4>
    <%
        for(int i=1;i <=9;i+ + ){
            for(int j=1;j<=i;j++){
    %>
    <%=i%>*<%=j%>=<%
        if(i*j <=9)
    %> 
    <%=i*j%>  
    <%
        if(j==i){
    %><br>
    <%
        }
            }
        }
    %>
  </body>
</html>
```

(2)查看 web.xml 的内容,部署项目,如图 9-5 所示。

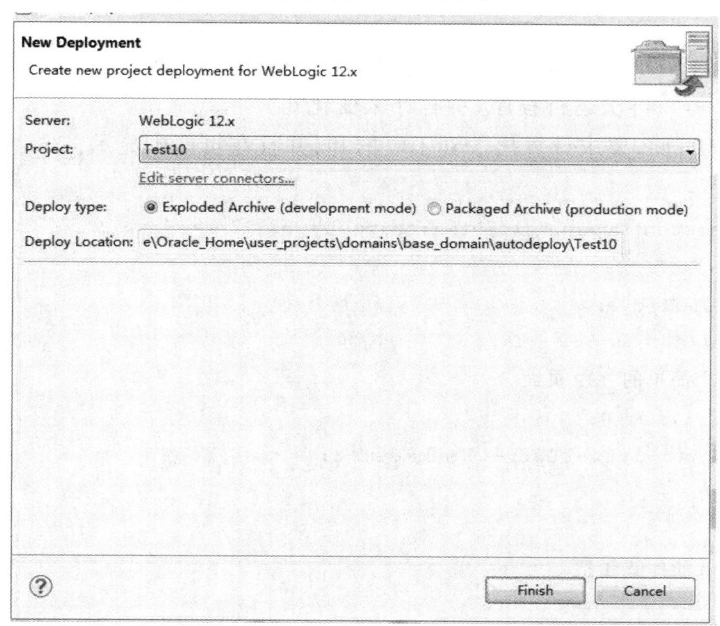

图 9-5　部署 Web 项目

(3)启动 WebLogic 服务器,打开浏览器,输入 http://localhost:7001/Test10/myjsp.jsp,观察浏览器输出的结果;不断刷新页面,观察显示结果有什么不同。

程序运行效果如图 9-6 所示。

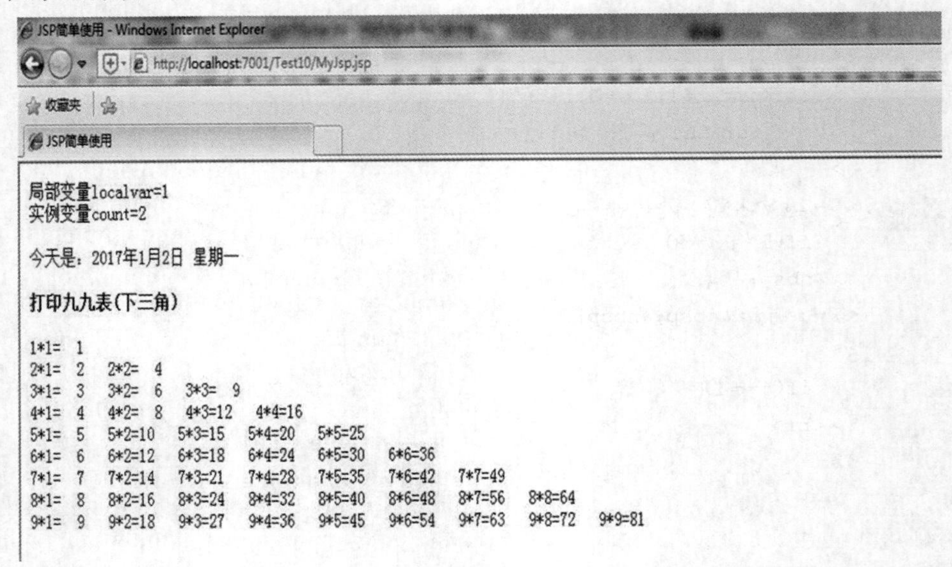

图 9-6 JSP 简单应用效果图

(4)回答下列问题：

①JSP 页面由哪些元素组成，各种元素在使用时，需要注意什么？

②观察带 JSP 的 Web 项目的配置文件 web.xml 与带 Servlet 的 Web 项目的配置文件有什么不同？

③直接修改 WebLogic 中 autodeploy 下 Web 项目中的 JSP 内容(如将输出内容加粗<b>…</b>等)，再次运行程序，看有什么变化？

2.编写 JSP 页面，要求计算出 1 到 100 的和，并且在页面中显示。

源代码如下：

```
<%@page contentType="text/html;charset=GB2312"%>
<HTML>
<BODY BGCOLOR=cyan>
<FONT Size=5>
<P>这是一个简单的 JSP 页面
    <%int i,sum=0;
       for(i=1;i<=100;i++){sum=sum+i;
       }
     %>
<P>1 到 100 的连续和是：
<BR>
   <%=sum %>
</FONT>
</BODY>
</HTML>
```

程序运行效果如图 9-7 所示。

图 9-7 使用 JSP 实现 1 到 100 的求和运算

3.编写 JSP 页面,要求可以在文本框中输入三条边的边长,然后判断三条边是否可以构建成一个三角形。如果可以则计算出三角形的面积,并且在页面中显示结果;如果不能则做出相应的提示。

源代码如下:

```
<%@page contentType="text/html;charset=GB2312" %>
<%@page import="java.util.* " %>
<HTML>
<BODY BGCOLOR=cyan><FONT Size=3>
  <P>请输入三角形的三个边的长度,输入的数字用逗号分隔:</p></FONT>
  <BR>
  <FORM action="" method=post name=form>
      <INPUT type="text" name="boy">
      <INPUT TYPE="submit" value="送出" name=submit>
  </FORM>
      <%! double a[]=new double[3];
          String answer=null;
      %>
      <%int i=0;
        boolean b=true;
        String s=null;
        double result=0;
        double a[]=new double[3];
        String answer=null;
        s=request.getParameter("boy");
        if(s!=null){
        // StringTokenizer 类的作用是按指定的字符串分拆字符串,本例中根据","分拆字符串
              StringTokenizer fenxi=new StringTokenizer(s,",");
              while(fenxi.hasMoreTokens()){
                  //逐个显示分拆后的字符串
                  String temp=fenxi.nextToken();
                  try{
```

```
                a[i]=Double.valueOf(temp).doubleValue();
                i++;
            }catch(NumberFormatException e){
                b= false;
                out.print("<BR>"+"请输入数字字符");
            }
        }
        if(a[0]+a[1]>a[2]&&a[0]+a[2]>a[1]&&a[1]+a[2]>a[0]&&b==true){
            double p=(a[0]+a[1]+a[2])/2;
            result=Math.sqrt(p*(p-a[0])*(p-a[1])*(p-a[2]));
            out.print("面积:"+result);
        }
        else{
            answer="您输入的三边不能构成一个三角形";
            out.print("<BR>"+answer);
        }
    }
%>
<P> 您输入的三边是:</p>
    <BR>
        <%=a[0]%>
    <BR>
        <%=a[1]%>
    <BR>
        <%=a[2]%>
</BODY>
</HTML>
```

程序运行效果如图 9-8 和图 9-9 所示。

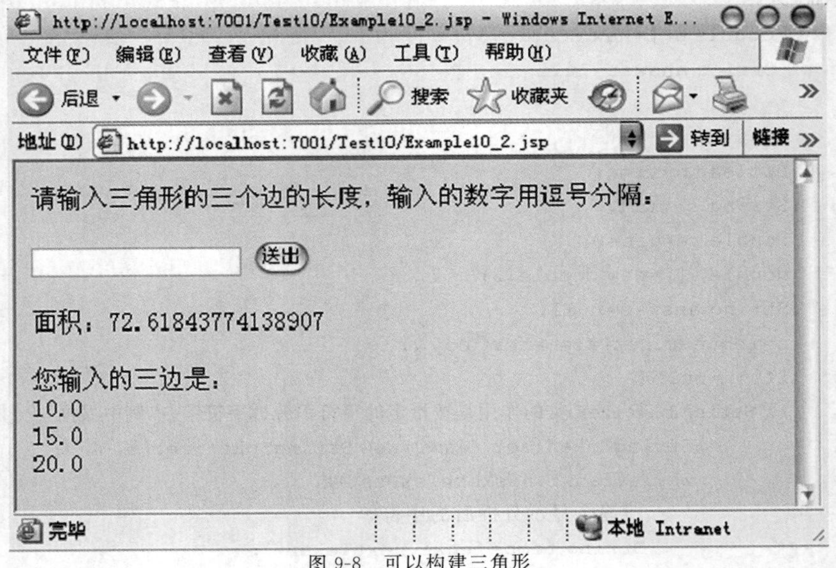

图 9-8　可以构建三角形

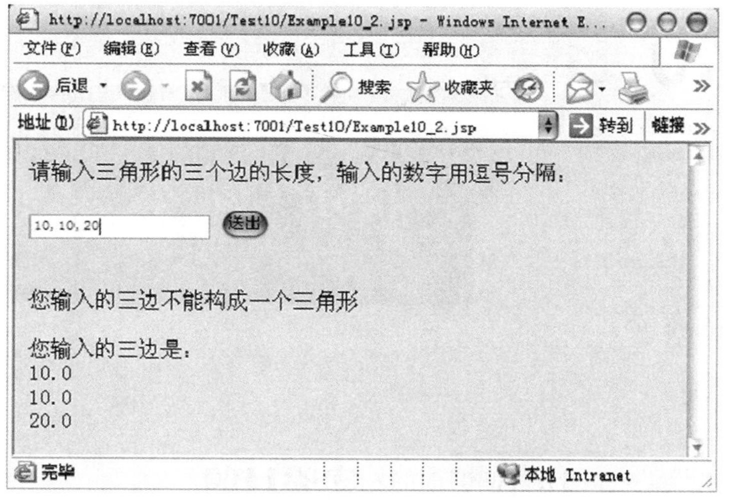

图 9-9　不能构建三角形

9.5 实验总结

本实验通过程序案例的形式让读者了解 JSP 页面的构成元素，介绍利用 MyEclipse 开发 JSP 的步骤，让读者掌握如何编写包含模板内容、指令、脚本元素、操作元素和注释等元素的 JSP 页面程序。在实验过程中，读者需要重点把握 JSP 与 Servlet 这两种技术的特点，体会这两种技术的优缺点，从而避免出现前面几个实验中使用 Servlet 实现 Web 程序的界面而导致 Web 项目显示层与业务逻辑层混淆不清的情况，为后续阶段高效开发 Web 项目打下基础。

9.6 课后思考题

1.简述 JSP 与 Servlet 的联系与区别。
2.JSP 页面由哪些元素组成，各种元素在使用时需要注意什么？
3.JSP 的执行过程包括哪两个阶段？
4.Web 容器将 JSP 转换为 Servlet 的程序位于哪个目录中？

实验 10 JSP 脚本及指令

10.1 实验目的

1.继续熟悉 JSP 基本元素模板、指令和脚本等的使用
2.熟悉在 JSP 中操作数据库的主要步骤
3.熟悉 JSP 中配置文件的使用
4.提高阅读、分析代码的能力

10.2 实验环境

1.MyEclipse 插件平台
2.WebLogic(或 Tomcat)容器

10.3 实验知识背景

10.3.1 JSP 的指令

1.JSP 指令概述

在 JSP 翻译成 Servlet 期间，指令提供整个 JSP 页面的相关信息，且不会产生任何输出信息到当前输出流中。JSP 指令有三类，分别为：page、include 和 taglib。

格式：

＜%@ 指令 属性名 1="…" 属性名 2="…" …%＞

或

＜jsp:directive.page 属性名 1="…" 属性名 2="…" …/＞

说明：＜与％、％与@、％与＞之间不能出现空格。

2.page 指令简介

page 指令是最复杂的 JSP 指令，其主要功能是设定整个 JSP 网页的属性和相关功能，具体功能由其属性来指定。page 指令的属性有 15 个，如表 10-1 所示。

表 10-1　　page 指令的属性

<table>
<tr><th>属　性</th><th>功　能</th></tr>
<tr><td>language="语言"</td><td>指定 JSP 容器要用什么语言来编译 JSP 网页。JSP 2.0 规范中指出，目前只可以使用 Java 语言</td></tr>
<tr><td>extends="基类名"</td><td>定义 JSP 网页转换成 Servlet 时继承的父类，通常不使用该属性</td></tr>
<tr><td>import="importList"</td><td>定义此 JSP 网页可以使用哪些 Java 类库，默认已导入四个包：java.lang.* ,java.servlet.* ,java.servlet.http.* ,java.servlet.jsp.* 。如果要导入多个包，既可以在一条语句中写完，也可以分多条语句来写，例如：
＜%@ page import="java.io. * "%＞
＜%@ page import="java.sql. * "%＞
与＜%@ page import="java.io. * ,java.sql. * "%＞等效</td></tr>
<tr><td>session="true|false"</td><td>指定此 JSP 网页是否可以使用 Session 对象。默认值为 true</td></tr>
<tr><td>buffer="none|size in kb"</td><td>指定输出流是否有缓冲区。默认值为 8 kB 的缓冲区</td></tr>
<tr><td>autoFlush="true|false"</td><td>指定输出流的缓冲区是否要自动清除，如果为 false，缓冲区满时会产生异常。默认值为 true</td></tr>
<tr><td>isThreadSafe="true|false"</td><td>指定此 JSP 网页是否能同时处理多个请求。默认值为 true，如果此值设为 false，JSP 在转换成 Servlet 时会实现 SingleThreadModel 接口</td></tr>
<tr><td>info="text"</td><td>指定此 JSP 网页的相关信息，可用 Servlet 接口的 getServletInfo()应用得到</td></tr>
<tr><td>errorPage="error_url"</td><td>如果发生异常或错误时，网页会被重新指向指定的 URL</td></tr>
<tr><td>isErrorPage="true|false"</td><td>表示此 JSP 网页是否为专门处理错误和异常的网页</td></tr>
<tr><td>contentType="ctinfo"</td><td>指定 MIME 类型和 JSP 网页的编码方式，其作用相当于 HttpServletResponse 接口的 setContentType()方法，例如：
＜%@ page contentType="text/html;charset=GB2312"%＞</td></tr>
<tr><td>pageEncoding="peinfo"</td><td>指定 JSP 页面的编码方式，如果设置了该属性，JSP 页面就以此方式编码，否则，就使用 contentType( )属性指定的字符集，假若两个属性都没有指定，则默认为"iso-8859-1"</td></tr>
<tr><td>isELIgnored="true|false"</td><td>指定 EL 表达式是否有效，如果为 true，则忽略 EL 表达式，否则，EL 表达式有效</td></tr>
<tr><td colspan="2">说明：还有两个属性不常用，此处不作介绍</td></tr>
</table>

3.include 指令

include 指令可以使 JSP 页面中静态包含一个文件，这个文件可以是 JSP 页面、HTML、文本文件或 Java 代码。当 JSP 转换成 Servlet 时，JSP 会在其中插入所包含文件的文本或代码，从而缓解代码冗余的问题。

格式：

＜%@ include file="相对于当前文件的 url"%＞

或

<jsp:directive.include file="相对于当前文件的 url"/>

4.taglib 指令

taglib 指令可以使 JSP 页面应用一些由 JSP 提供的动作元素标签去完成特定的功能，如使用<jsp:include>标签可以在 JSP 页面中包含一个文件。另外，taglib 指令还允许开发人员使用自定义标签来完成特定的功能。

格式：

<%@ taglib uri="tagURI" prefix="tagPrefix"%>

说明：

uri 属性：表示唯一标识标签库描述符(TLD)的 uri，在标签库描述符中描述了 uri。这个 uri 可以是直接或者非直接的。

prefix 属性：定义了区分指定标签库所定义的标签与其他标签库提供的标签的前缀。

10.3.2 JSP 的脚本元素

JSP 脚本元素主要用于在页面中嵌入 Java 代码，以实现页面的动态请求。JSP 的脚本元素主要由小脚本、表达式和声明组成。

1.JSP 小脚本

JSP 小脚本就是在 JSP 页面里嵌入一段 Java 代码。

语法：

<% Java 代码 %>

例如，下面这段代码是使用 JSP 小脚本把系统时间显示在页面上，效果如图 10-1 所示。

```
<%@page language="java" import="java.util.*,java.text.*"
    contentType="text/html;charset=GBK"%>
<html>
    <head><title>输出当前日期</title></head>
    <body>
    你好,今天是
    <%
        SimpleDateFormat formater=new SimpleDateFormat("yyyy年MM月dd日");
        String strCurrentTime=formater.format(new Date());
        out.print(strCurrentTime);
    %>
    </body>
</html>
```

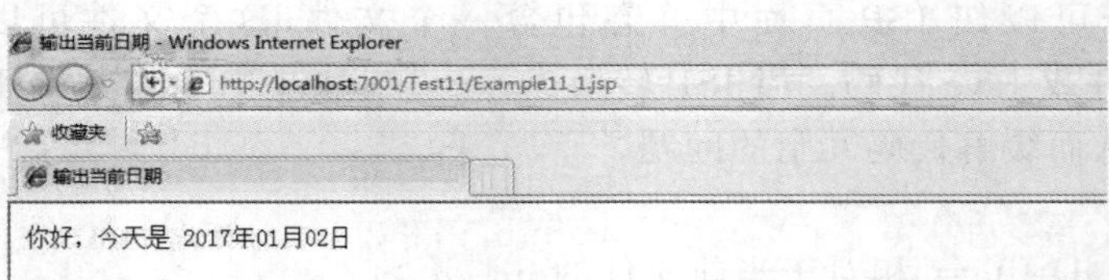

图 10-1　使用 JSP 小脚本显示当前系统时间

2.JSP 表达式

JSP 表达式是对数据的表示，系统将其作为一个值进行计算和显示。

语法：

＜％＝Java 表达式/变量％＞

例如，使用下面小脚本显示乘积：

```
<html>
<%
    out.println(30*20);
%>
</html>
```

效果与使用下面例子中 JSP 表达式是一样的：

```
<html>
<%=30*20%>
</html>
```

显然，在 HTML 中显示数据，使用表达式会更方便。

3.JSP 声明

在 JSP 文件中，同样可以像 Java 类一样在页面内部声明方法，以解决代码冗余的问题。

语法：

＜％！Java 代码 ％＞

例如，在同一个 JSP 页面中，如果需要在多个地方格式化日期，可以通过在 JSP 中声明一个方法来按照特定日期格式对输出内容进行排版，页面效果如图 10-2 所示。

```
<%@page language="java"import="java.util.*,java.text.*"contentType="text/html;
charset=GBK"%>
<html>
<%!
    String formatDate(Date d){
        SimpleDateFormat formater=new SimpleDateFormat("yyyy 年 MM 月 dd 日");
        return formater.format(d);
    }
%>
你好,张三!今天是<%=formatDate(new Date())%><br>
你好,李四!今天是<%=formatDate(new Date())%>
</html>
```

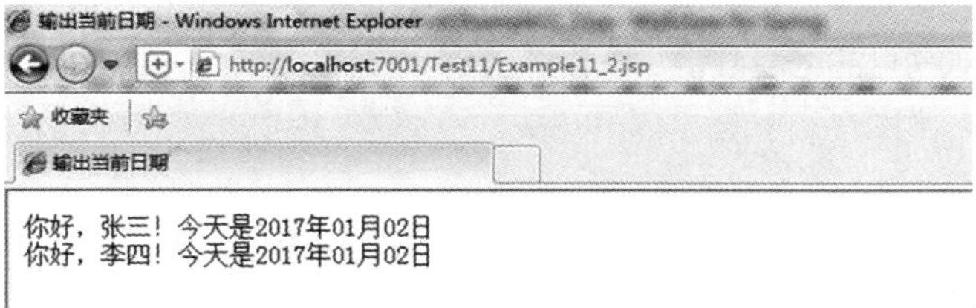

图 10-2　使用 JSP 声明方法显示当前系统时间

10.4 实验内容与步骤

1.编写一个 JSP 页面，实现根据一个人 18 位身份证号码显示生日的功能，要求用到表达式、方法声明和小脚本。

源代码：

```
<%@page language="java" contentType="text/html;charset=GBK"%>
<%!
    final String separator="-";
    public String getBirthday(String identify){
      String birthday="";
      birthday=identify.substring(6,10)+separator+identify.substring(10,12)
      +separator+identify.substring(12,14);
      return birthday;
    }
%>
<%
  String identify1="010020198810092211";
  String birthday1=getBirthday(identify1);
  String identify2="010020199009302211";
  String birthday2=getBirthday(identify2);
%>
<html>
  <head>
    <title>计算生日</title>
  </head>
  <body bgcolor="# ffffff">
  <h4 align="center">根据身份证自动计算生日 </h4>
  <table width="400" border="1" align="center">
    <tr>
        <td>
          <div align="center">身份证</div>
        </td>
        <td>
          <div align="center">生日</div>
        </td>
      </tr>
      <tr>
        <td> <%=identify1% ></td>
        <td> <%=birthday1% ></td>
      </tr>
```

```
      <tr>
        <td> <%=identify2%></td>
        <td> <%=birthday2%></td>
      </tr>
    </table>
      <h4 align="center"> </h4>
    <br/>
  </body>
</html>
```

程序运行效果如图 10-3 所示。

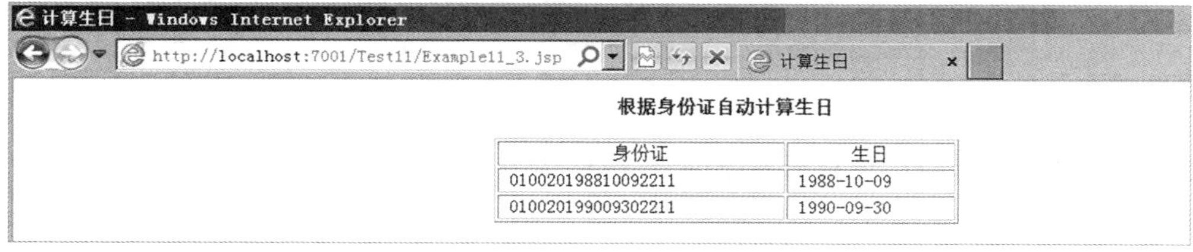

图 10-3　使用 JSP 提取身份证上的生日

2.按要求编写 JSP 程序，实现从数据库读取学生的学号、姓名、语文和数学成绩等信息，在 JSP 页面上显示。

(1)首先使用 MySQL 建立数据库 mydb，在数据库中建立数据表 stdb，如图 10-4 所示。

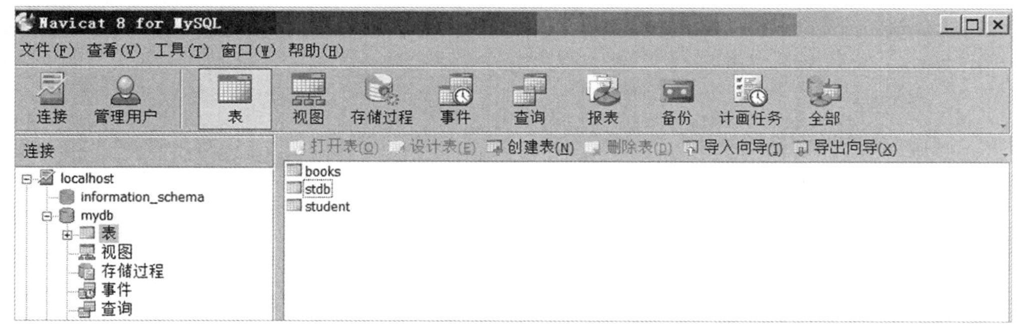

图 10-4　建立数据库

(2)数据表 stdb 结构如图 10-5 所示，建立表后向表中输入若干记录。

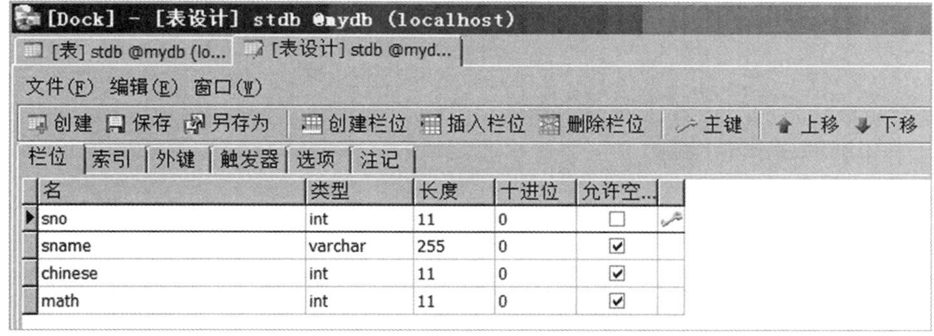

名	类型	长度	十进位	允许空...
sno	int	11	0	☐
sname	varchar	255	0	☑
chinese	int	11	0	☑
math	int	11	0	☑

图 10-5　建立数据表

(3)在 WebLogic Console 中建立上述数据库的 JDBC 数据源(Data Source),配置该数据源的 JNDI 名字为“AA”,如图 10-6 所示。

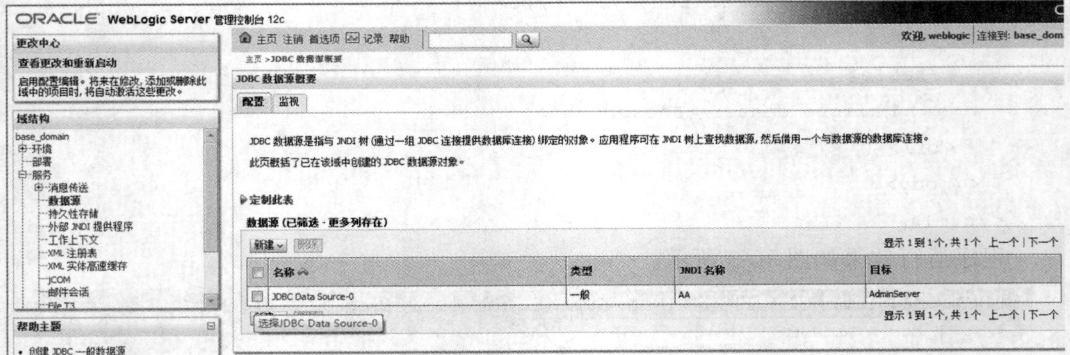

图 10-6　配置 WebLogic 的数据源与 JNDI 服务

(4)新建 Web 项目,项目中新建 JSP 文件。

源代码如下:

```
<%@page contentType="text/html;charset=GB2312"%>
<%@page import="java.sql.*"% >
<%@page import="java.util.*"% >
<%@page import="javax.naming.*"% >
<%@page import="javax.sql.*"% >
<%@page import="java.io.*"% >
<HTML>
    <BODY>
        <table border="1">
            <tr>
                <th>
                    学号
                </th>
                <th>
                    姓名
                </th>
                <th>
                    语文
                </th>
                <th>
                    数学
                </th>
            </tr>
            <%
                DataSource ds=null;
        Context ctx;
```

```
        Connection myConn=null;
        final String JNDI_DATABASE_NAME="AA";
        Properties ht=new Properties();
        ht.put(Context.INITIAL_CONTEXT_FACTORY,"weblogic.jndi.WLInitialContextFactory");
        ht.put(Context.PROVIDER_URL,"t3://localhost:7001");
        try{
            ctx=new InitialContext(ht);
            ds=(javax.sql.DataSource)ctx.lookup(JNDI_DATABASE_NAME);
        }catch(Exception e){
            e.printStackTrace();
        }
        if(ds==null){
            System.out.println("Eorror !");
        }
        else{
            System.out.println("Connection is OK !");
        }
        PreparedStatement myStatement=null;
        ResultSet mySet=null;
        try{
            myConn=ds.getConnection();
            myStatement=myConn.prepareStatement("select* from stdb");
            mySet=myStatement.executeQuery();
            while(mySet.next()){
        %>
            <tr>
                <td><%=Integer.toString(mySet.getInt("sno"))%></td>
                <td><%=mySet.getString("sname")%></td>
                <td><%=Integer.toString(mySet.getInt("chinese"))%></td>
                <td><%=Integer.toString(mySet.getInt("math"))%></td>
            </tr>
<%
            }
            myStatement.close();
            mySet.close();
            myConn.close();
        }catch(Exception e){
            e.printStackTrace();
        }
%>
        </table>
```

```
</BODY>
</HTML>
```

(5)部署 Web 项目,在浏览器中输入网址,观察浏览器输出的结果。

程序运行效果如图 10-7 所示。

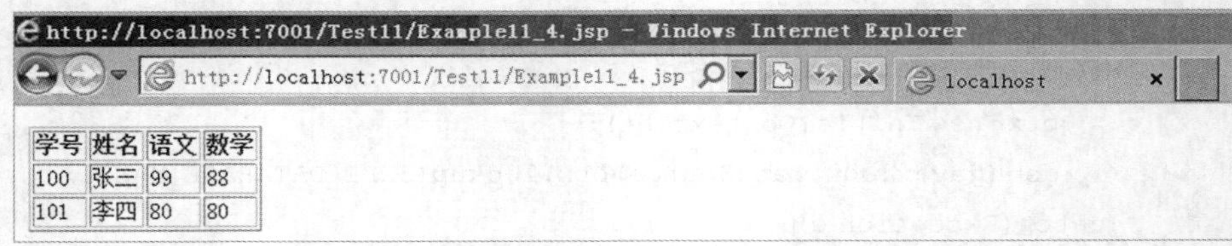

学号	姓名	语文	数学
100	张三	99	88
101	李四	80	80

图 10-7　使用 JSP 读取数据库

3.按要求编写 JSP 程序,实现从 web.xml 配置文件中读取预设的密码信息进行登录验证,验证通过后可以进行表达式运算并且将运算结果进行显示。

(1)建立 A.jsp 页面,页面上设置文本输入框控件,使用 JavaScript 技术对输入的密码进行输入验证,如果输入非数字的密码,则进行相应提示并且清除密码内容请求再次输入。

源代码如下:

```
<%@page language="java" import="java.util.*" pageEncoding="UTF-8"%>
<html>
    <head>
        <title>A.jsp</title>
<script language="javascript" type="text/javascript">
    function CheckMyForm(){
        var txt=document.myform.pass;
        if(!checkNumber(txt)){
            return false;
        }
        return true;
    }
    //验证只能为数字
    function checkNumber(obj){
    var reg=/^[0-9]+$/;  //正则表达式
    if(!reg.test(obj.value)){
        alert('只能输入数字!');
        obj.value="";
        obj.focus();
        return false;
    }
     return true;
}
</script>
    </head>
```

```
        <h1>
            请登录系统
        </h1>
        <body>
            <FORM action="B.jsp" method="post" name="myform"
                onsubmit="return CheckMyForm()">
                <INPUT type="password" name="pass">
                <INPUT TYPE="submit" value="登录" name=submit>
            </FORM>
        </body>
</html>
```

程序运行效果如图 10-8 和图 10-9 所示。

图 10-8　登录密码输入界面

图 10-9　输入非数字密码后的提示

(2)当单击【登录】按钮后进入 B.jsp 页面,实现登录密码验证。首先在 web.xml 中为 B.jsp 配置初始化参数,参数名为“secret”,值为“123456”。

源代码如下:

```
<?xml version="1.0" encoding="UTF-8"?>
<web-app version="2.5" xmlns="http://java.sun.com/xml/ns/javaee"
xmlns:xsi="http://www.w3.org/2001/XMLSchema-instance" xsi:schemaLocation=
"http://java.sun.com/xml/ns/javaee  http://java.sun.com/xml/ns/javaee/web-
app_2_5.xsd">
<servlet>
  <servlet-name>myservlets</servlet-name>
  <jsp-file>/B.jsp</jsp-file>
  <init-param>
   <param-name>secret</param-name>
   <param-value>123456</param-value>
  </init-param>
</servlet>
<servlet-mapping>
  <servlet-name>myservlets</servlet-name>
  <url-pattern>/B.jsp</url-pattern>
</servlet-mapping>
```

```
<welcome-file-list>
  <welcome-file>A.jsp</welcome-file>
</welcome-file-list>
<login-config>
  <auth-method>BASIC</auth-method>
</login-config>
</web-app>
```

判断用户在输入框中输入的密码与“secret”的值是否相等，若不等，跳转到错误页面error.jsp。

源代码如下：

```
<%@page language="java" import="java.util.*" pageEncoding="UTF-8"%>
<html>
  <head>
    <title>B.jsp</title>
  </head>
  <%--重写 jspInit() --%>
  <%!
     String str=null;
     public void jspInit(){
        ServletConfig sc=getServletConfig();
        str=sc.getInitParameter("secret");
     }
   %>
  <body bgcolor="#ffffff">
  <h1>计算器</h1>
    <%
      if(!str.equals(request.getParameter("pass"))){
         RequestDispatcher view=request.getRequestDispatcher("error.jsp");
         view.forward(request,response);
        // view.include(request,response);
      }
    %>
       <FORM action="C.jsp" method="post" name="myform"
          onsubmit="return CheckMyForm()">
          <INPUT type="text" name="expr">
          <INPUT TYPE="submit" value="计算" name="submit">
       </FORM>
  </body>
</html>
```

程序运行效果如图 10-10 和图 10-11 所示。

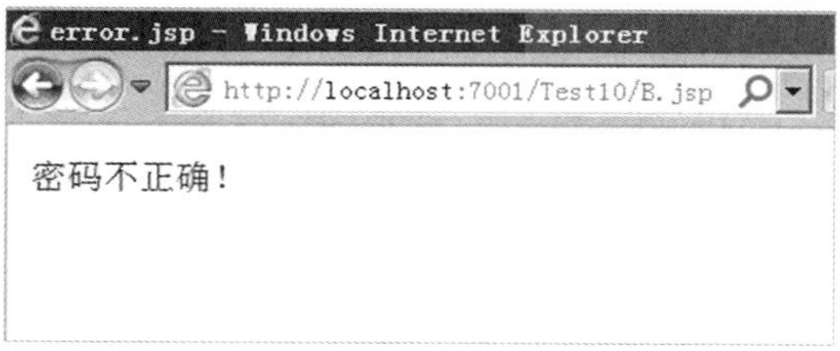

图 10-10　输入的密码不是“123456”

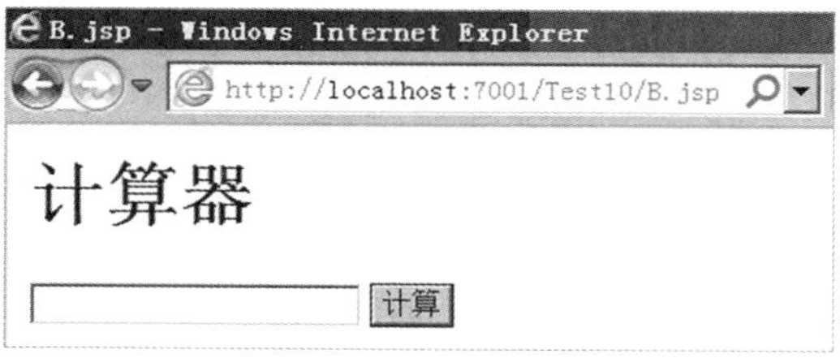

图 10-11　正确输入密码“123456”

(3)要实现字符串运算符的计算，项目需要导入如图 10-12 所示的 jar 包。

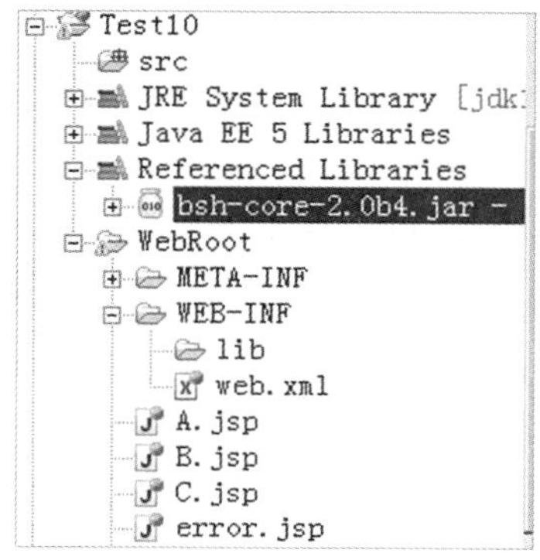

图 10-12　导入 bsh-core-2.0b4.jar

当在 B.jsp 正确输入运算表达式后，进入页面 C.jsp，求出 B.jsp 中输入的运算字符串的结果并且显示。

源代码如下：

```
<%@page language="java" import="java.util.*,bsh.*" pageEncoding="utf-8"% >
<html>
  <head>
      <title>C.jsp</title>
  </head>
  <body>
  <h1>计算结果</h1>
  <%
     Interpreter bsh=new Interpreter();
        try{
            bsh.eval("el="+request.getParameter("expr"));
            out.println(request.getParameter("expr") + "=");
            out.println(bsh.get("el"));
        }catch(Exception e){}
   %>
  </body>
</html>
```

程序运行效果如图 10-13 和图 10-14 所示。

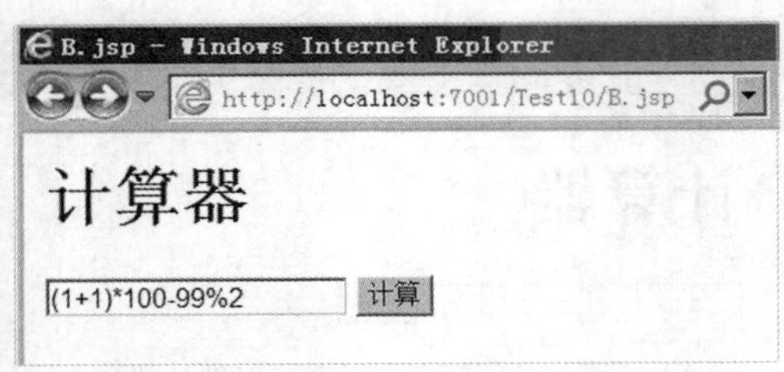

图 10-13　输入运算表达式

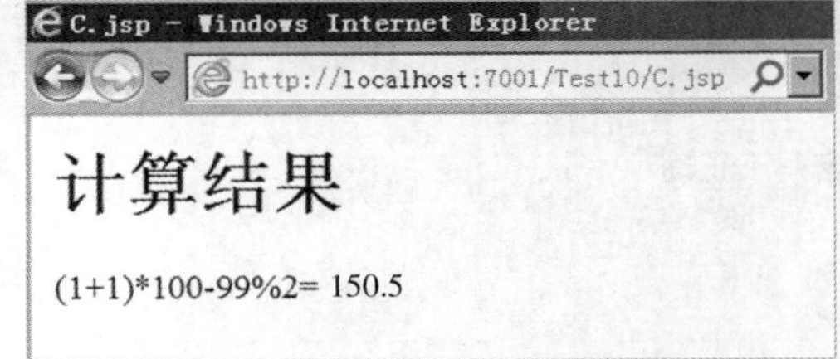

图 10-14　显示表达式的计算结果

10.5 实验总结

本实验让读者进一步熟悉 JSP 脚本与指令的使用，JSP 指令一共有 3 个，包括 page 指令、include 指令和 taglib 指令。page 指令是新建一个 JSP 页面必须用到的指令，page 指令包含描述该页面使用的编码、该页面所使用的包以及使用什么样的语言作为脚本语言等。JSP 脚本的实验重点介绍了小脚本、表达式与声明方法的操作步骤，为下一阶段学习和使用 EL 表达式及 JSTL 标签打下基础。

10.6 课后思考题

1.page 指令、include 指令和 taglib 指令的作用分别是什么？
2.使用＜%!%＞声明方法和使用 JavaScript 有什么异同？
3.怎样在小脚本中访问某个类的实例成员？
4.表达式能否以“;”结尾？

实验 11 JSP 隐式对象

11.1 实验目的

1.理解 JSP 隐式对象的相关概念
2.理解 JSP 中隐式对象的作用范围
3.掌握在 JSP 页面中使用隐式对象来实现相关操作的方法
4.提高阅读、分析代码的能力

11.2 实验环境

1.MyEclipse 插件平台
2.WebLogic(或 Tomcat)容器

11.3 实验知识背景

11.3.1 JSP 隐式对象的介绍

1.JSP 隐式对象概述

隐式对象是一种在 JSP 页面的文件中不用声明就可以使用的对象。JSP 隐式对象又称为内置对象或内部对象。JSP 为简化页面的开发,向开发人员提供了一些隐式对象,这些对象实际上是 Web 容器加载的一组类的实例,但是不需要像一般的 Java 实例对象那样用"new"关键字来获取实例,而是一种可以直接在 JSP 页面使用的对象。所有的隐式对象只对 Scriptlet 或者表达式起作用。隐式对象使用的类所对应的包为 javax.servlet.jsp.*、javax.servlet.*和 javax.servlet.http.*。

2.隐式对象的分类

(1)输入和输出对象

输入和输出对象控制页面的输入和输出,包括 request、response 和 out。

(2)作用域通信对象

作用域通信对象检索与 JSP 页面的 Servlet 相关的信息,包括 session、application 和 pageContext。

(3)Servlet 对象

Servlet 对象提供有关页面环境的信息,包括 config 和 page。

(4)错误对象

错误对象用来处理 JSP 页面中的错误,包括 exception。

常用的 JSP 隐式对象有 9 种,如表 11-1 所示。

表 11-1 9 种 JSP 隐式对象

隐式对象	类 型	说 明
request	javax.servlet.http.HttpServletRequest	隐含请求信息
session	javax.servlet.HttpSession	表示会话对象
application	javax.servlet.ServletContext	JSP 页面所在 Web 应用的上下文对象
response	javax.servlet.HttpServletResponse	响应信息
out	javax.servlet.JspWriter	JSP 的数据输出对象
pageContext	javax.servlet.jsp.PageContext	当前 JSP 页面的上下文对象
page	java.lang.Object	对当前 JSP 页面的引用,即 Java 中的 this
config	javax.servlet.ServletConfig	JSP 页面的 ServletConfig 对象
exception	java.lang.Throwable	异常处理

11.3.2 out 输出对象

out 对象表示一个输出流,用来向客户端输出数据,也就是在浏览器内输出信息。

out 输出对象的常用方法有:

(1)out.print(boolean)　　out.println(boolean)

(2)out.print(char)　　out.println(char)

(3)out.print(double)　　out.println(double)

(4)out.print(float)　　out.println(float)

(5)out.print(long)　　out.println(long)

(6)out.print(String)　　out.println(String)

(7)out.newLine():输出一个换行符

(8)out.flush():输出缓冲区里的内容

(9)out.close():关闭流

在使用这些方法时要注意:out.println(参数)在后面加一个换行符(不是指显示结果换行),而 out.print(参数)不会在输出数据后自动换行,例如:

```
<%
    out.println("Test");
```

```
    out.println("Test");
    out.print("Test");
    out.print("Test");
%>
```

这段 JSP 小脚本运行的结果如图 11-1 所示。

Test Test TestTest

图 11-1 out.println(参数)与 out.print(参数)显示效果对比

以下是使用 out 对象向客户输出信息(包括表格等内容)的示例:

```
<%
    out.print("<Table width=540 border=1 bordercolor=0000ff cellspacing=0
    cellpadding=0 height=53>");
    out.print("<CAPTION ><H1>学生成绩登记表</H1></CAPTION>");
    out.print("<TR align=center >");
    out.print("<TH width=120 height=27>序号</TH>");
    out.print("<TH width=120 height=27>学号</TH>");
    out.print("<TH width=80 height=27>姓名</TH>");
    out.print("<TH width=120 height=27>平时成绩</TH>");
    out.print("<TH width=120 height=27>考试成绩</TH>");
    out.print("<TH width=120 height=27>总评成绩</TH>");
    out.print("</TR>");
    out.print("<TR align=center >");
    out.print("<TD width=120 height=22>1</TD>");
    out.print("<TD width=120 height=22>0378122654</TD>");
    out.print("<TD width=80 height=22>王二</TD>");
    out.print("<TD width=120 height=22>78</TD>");
    out.print("<TD width=120 height=22>87</TD>");
    out.print("<TD width=120 height=22>84</TD>");
    out.print("</TR>");
    out.print("</Table>");
%>
```

运行效果如图 11-2 所示。

学生成绩登记表

序号	学号	姓名	平时成绩	考试成绩	总评成绩
1	0378122654	王二	78	87	84

图 11-2 使用 out 对象输出表格

11.3.3 request 请求对象

1.request 对象介绍

request 对象用来封装用户提交的信息。该对象调用相应的方法,则可获取封装信息，如请求参数、Cookie、HTTP 请求头以及客户端 IP 地址等。

request 对象的常用方法有：

(1)getProtocol()：获取客户向服务器提交信息所使用的通信协议，如 HTTP1.1 等。

(2)getServletPath()：获取客户请求的 JSP 页面文件的目录(相对于根目录的地址)。

(3)getContentLength()：获取客户提交的整个信息的长度。

(4)getMethod()：获取客户提交信息的方式，如 post 或 get。

(5)getHeader(String s)：获取 HTTP 头文件中由参数 s 指定的头名字的值。s 参数有：accept、referer、accept-language、content-type、accept-encoding、use-agent、host、content-length、connection、cookie 等，如 getHeader("user-agent ")表示获取客户的浏览器版本号等信息。

(6)getHeaderNames()：获取头名字的一个枚举。

(7)getHeaders(String s)：获取头文件中指定头名字的全部值的一个枚举。

(8)getRemoteAddr()：获取客户端的 IP 地址。

(9)getRemoteHost()：获取客户机的名称(如果获取不到，就获取 IP 地址)。

(10)getServerName()：获取服务器的名字。

(11)getServerPort()：获取服务器的端口号。

(12)getParameter(String name)：根据 name 获取传入的参数。

(13)getParameterNames()：获取传入参数的名字集合。

(14)getLocale()：获取本地的国家和语言。

以下是使用 request 隐式对象获取客户提交的信息示例：

```
<%--提交请求的 JSP 页面--%>
<FORM action="addOper.jsp" method=post>
      <INPUT type="text" name="num1" size="8" value="1">
      <INPUT TYPE="submit" value="提交" >
</FORM>
<%--获取请求中客户提交信息的 JSP 页面--%>
<P>您的输入是：
   <%String textContent1=request.getParameter("num1");%>
<BR>
<%=textContent1%>
```

2.request 对象的使用要点

(1)使用 request 对象获取信息时很容易出现异常，例如：

```
String textContent=request.getParameter("data");
double number=Double.parseDouble(textContent);
```

这两条语句结合起来使用，很容易出现 NullPointerException 异常。因为当 data 文本框没有输入任何数据时，textContent 也就为 null，而 Double.parseDouble(String s)方法要求 s 不为 null，否则就会抛出 NullPointerException 异常。为了避免在运行时容易出现空对象异常，一般做如下处理：

```
String textContent=request.getParameter("data");
```

```
if(textContent==null){
    textContent="0.0";
}
double number=Double.parseDouble(textContent);
```

或者

```
String textContent=request.getParameter("data")==null?"0.0":request.getParameter
("data");
double number=Double.parseDouble(textContent);
```

(2)当用 request 对象获取客户提交的汉字字符时,无法正确显示。通常在使用 request 之前,根据表单提交请求的方式做如下处理即可解决中文汉字的显示问题。

①Form 表单的 method 方式是 POST 的情况:

```
<%@page contentType="text/html;charset=GBK" %>
request.setCharacterEncoding("GBK");
```

②Form 表单的 method 方式是 GET 的情况,需要对参数解码,然后重新编码。例如:

```
<%@page contentType="text/html;charset=GBK" %>
String info=request.getParameter("info");
info=new String(info.getBytes("ISO8859_1"),"GBK");
```

11.3.4 response 响应对象

1.response 对象简介

response 对象用于对客户的请求做出动态响应,向客户端发送数据。例如:根据客户要求,设置客户端输出信息的字符集,重新定向客户端的请求,或者添加 Cookie 等。在 JSP 文件中该对象比较少用,因为该对象属性的主要功能由 page 标签完成。

response 对象的常用方法有:

(1)setContentType(String s):改变 ContentType 的属性值。

(2)add[set]Header(String head,String value):动态添加新的响应头和头的值。

(3)sendRedirect(String location):重新定向客户端的请求。

(4)setStatus(int n):设置响应的状态行。

(5)setLocale(Locale loc):设置本地的国家和语言。

(6)addCookie(Cookie cookie):加入 Cookie 方法。

2.response 对象的应用

(1)动态响应 ContentType 属性

由于 page 指令只能为 ContentType 指定一个值,用以决定响应的 MIME(内容)类型,如果想动态地改变这个属性的值来响应客户端,就需要使用 setContentType(String s)方法来改变 ContentType 的属性值,格式如下:

```
public void setContentType(String s);
```

其中,参数 s 的 MIME 类型部分可取 text/html text/plain(以文本类型提交给客户)/application/x-msexcel application/msword 等。

(2)重定向

在某些情况下，当响应客户端时，需要将客户端重新引导至另一个页面，可以使用sendRedirect(URL url)方法实现客户的重定向。例如：

```
response.sendRedirect("http://www.sise.com.cn");
```

11.3.5 session 会话对象

session 会话对象可以记录客户端的访问状态。从一个客户打开浏览器并连接到服务器的时刻开始，到客户关闭浏览器离开这个服务器的时刻结束，称为一个会话。保存在该对象中的变量值可以在客户端访问的整个周期内使用。每一个 session 对象由服务器分配一个专属的 ID 以区别身份，session ID 的工作原理如图 11-3 所示。

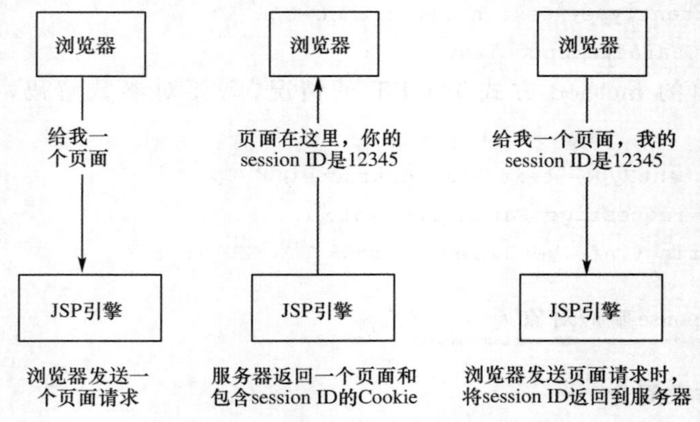

图 11-3 session ID 的工作原理示意图

session 对象的常用方法有：

1.public void setAttribute(String key,Object obj)

将对象 obj 添加到 session 对象中，并为添加的对象指定一个索引关键字。如果添加的两个对象的关键字相同，则先添加的对象被清除。

2.public Object getAttribute(String key)

获取 session 对象中关键字是 key 的对象。该方法返回的对象，应强制转换为原来的类型。例如：

String 顾客=(String)session.getAttribute("customer");

3.public Enumeration getAttributeNames()

产生一个枚举对象，该枚举对象使用 nextElements()方法遍历 session 对象中所含有的全部对象。

4.public void removeAttribute(String key)

从当前 session 对象中删除关键字是 key 的对象。

5.public String getId()

获取 session 对象的编号。

6.public long setMaxInactiveInterval(int n)

设置 session 对象的生存时间(单位是秒)。

7.public long getMaxInactiveInterval()

取得 session 对象的生存时间。

8.public boolean isNew()

判断该用户是否是新用户。

下面是使用 session 对象实现一个网站访问计数器的示例：

```
<%! public int number=0;
    synchronized void countPeople(){number++;}
%>
<%if(session.isNew()){
    countPeople();
    String str=String.valueOf(number);
    session.setAttribute("count",str);
}%>
<P>
您是第<%=(String)session.getAttribute("count")%>个访问本站的人。
```

11.3.6 application 对象

服务器启动后，就会产生 application 对象。当一个客户访问服务器上的一个 JSP 页面时，JSP 引擎为该客户分配这个 application 对象，当客户在所访问的网站的各个页面之间浏览时，都使用同一个 application 对象，直到服务器关闭，这个 application 对象才被取消。保存在该对象中的变量值可以让所有客户端使用。与 session 对象不同的是，所有客户都共享同一个 application 对象。

application 对象的常用方法有：

1.setAttribute(String key,Object obj)

2.getAttribute(String key)

3.removeAttribute(String key)

11.3.7 pageContext 对象

pageContext 对象使用户可以访问页面作用域中定义的所有隐式对象。pageContext 对象所提供的方法可以访问隐式对象在页面上定义的所有属性。它的作用范围仅在页面内。

pageContext 对象的常用方法有：

(1)getRequest():返回当前的 request 对象

(2)getResponse():返回当前的 response 对象

(3)getSession():返回当前的 session 对象

(4)getOut():返回输出对象

(5)getException():返回当前的 exception 对象

(6)getPage():返回当前页面对象

(7)getServletConfig():返回当前页面的 ServletConfig 对象

(8)getServletContext():返回 ServletContext 对象。这个对象对所有页面都是共享的。

pageContext 对象也提供存取属性的方法:

(1)void setAttribute(String name,Object value)

(2)Object getAttribute(String name)

(3)void removeAttribute(String name)

使用时,还可以带上指定范围的参数:

①Object getAttribute(String name,int scope)

②void removeAttribute(String name,int scope)

③void setAttribute(String name,Object value,int scope)

范围参数的值是指 pageContext 的四个常量:PAGE_SCOPE、REQUEST_SCOPE、SESSION_SCOPE、APPLICATION_SCOPE,具体范围如图 11-4 所示。

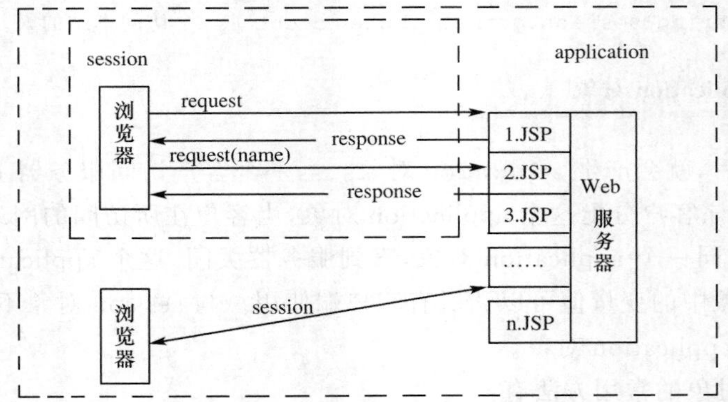

图 11-4 request、response、session、application 作用域的比较

(4)Object findAttribute(String name):按 page、request、session、application 顺序查找属性。

(5)void include(String relativeUrlPath):将请求转发给参数 Url 定义位置的对象,被调用的对象对请求做出响应将并入原先的响应对象中。

(6)void forward(String relativeUrlPath):将请求转发给参数 Url 定义位置的对象,被调用的对象对请求做出响应,原对象的响应被中止。与 include()方法执行效果的区别如图 11-5 所示。

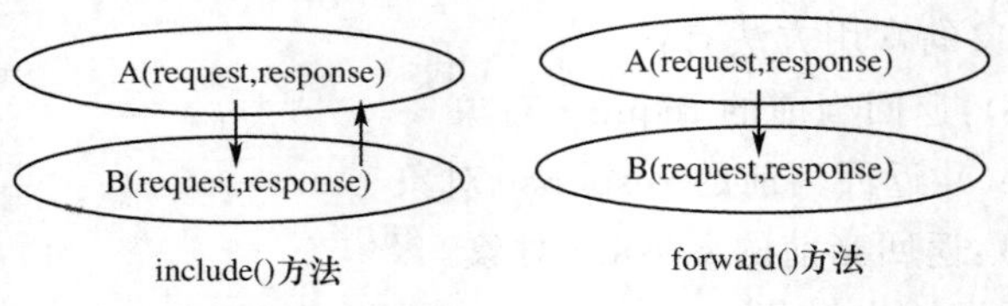

图 11-5 include()方法与 forward()方法的对比

11.4 实验内容与步骤

结合前面章节，使用 JSP 技术实现一个可以进行在线课堂测试的 Web 项目。

1.项目的业务流如图 11-6 所示。

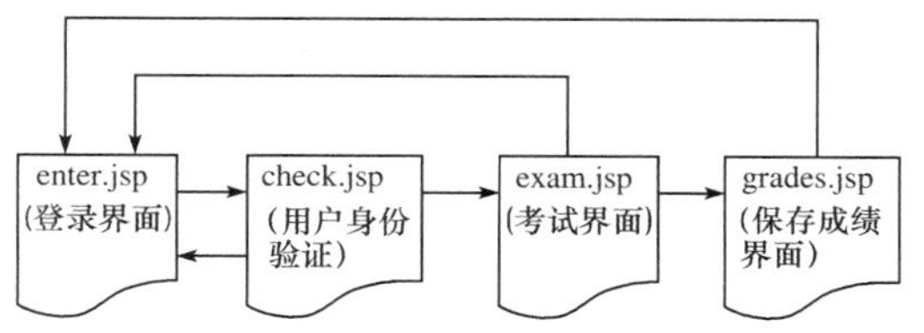

图 11-6 在线课堂测试系统的业务流

2.使用 Access 建立数据库，在数据库中建立一张表，命名为 students，用于保存学生信息及成绩。表结构如图 11-7 所示。

字段名称	数据类型
id	文本
name	文本
password	文本
classid	文本
score1	数字
score2	数字
score3	数字
score4	数字
score5	数字
ip1	文本
1p2	文本

字段属

图 11-7 students 表结构

3.创建一个 Excel 文件，命名为 test1.xls，并设置 3 张页面，分别用来保存某一次测试的题目及答案(题型为判断题、单选题和多选题)，样式如图 11-8 所示。

	A	B	C
1	question	key	
2	AWT是重量级组件，显示图形界面时不依赖于操作系统的对等组件。	F	
3	Frame窗体的默认布局管理器是BorderLayout。	T	
4	进行AWT界面设计时，需要导入java.awt包中的类和接口。	T	
5	FlowLayout的默认对齐方式是左对齐。	F	
	在AWT组件的继承关系中，Container是Component的		

judges / chooses /

A	B	C	D	E	F
question	A	B	C	D	key
向容器添加组件的方法是【 】	add( )	append( )	addItem( )	insert( )	A
面板(Panel)默认的布局管理器是【 】	BorderLayout	FlowLayout	GridLayout	CardLayout	B
以下哪个布局管理器可以用于国际象棋游戏应用程序？【 】	BorderLayout	FlowLayout	GridLayout	CardLayout	C
下列组件中，属于中间容器的是【 】	Frame	Container	Panel	Button	C

A	B	C	D	E	F
question	A	B	C	D	key
下列类中，能够显示图形形状、大小的是【 】	Frame	Jframe	Panel	Applet	ABD

图 11-8 test1.xls 样式

4.将 Access 文件 students.mdb 和 Excel 文件 test1.xls 存放到学生不能直接访问的位置(如 F:\ejava_db 中)，具体位置在 web.xml 中指定。若放在 WebRoot 目录中，学生能够下载，会导致试题泄密。

5.在 web.xml 中定义项目需要用到的初始化参数。

web.xml 主要内容：

```
<context-param>
    <param-name>accessfile</param-name>
    <param-alue>f:/ejava_db/students.mdb</param-value>
</context-param>
<context-param>
    <param-name>excelfile</param-name>
    <param-value>f:/ejava_db/test1.xls</param-value>
</context-param>
<context-param>
    <param-name>ip</param-name>
    <param-value>ip1</param-value>
</context-param>
<context-param>
    <param-name>score</param-name>
    <param-value>score1</param-value>
</context-param>
<context-param>
    <param-name>value1</param-name>
    <param-value>5</param-value>
</context-param>
<context-param>
    <param-name>value2</param-name>
    <param-value>5</param-value>
</context-param>
<context-param>
    <param-name>value3</param-name>
    <param-value>10</param-value>
</context-param>
<error-page>
    <error-code>404</error-code>
    <location>/exam/404.jsp</location>
</error-page>
<error-page>
    <error-code>500</error-code>
    <location>/exam/500.jsp</location>
</error-page>
<welcome-file-list>
    <welcome-file>enter.jsp</welcome-file>
    </welcome-file-list>
```

①指定Access文件位置

②指定excel文件位置

③记录考生ip的字段

④记录考生成绩的字

⑤设置判断题分值

⑥设置单选题分值

⑦设置多选题分值

⑧指定404错误处理程序

⑨指定500错误处理程序

⑩默认文件

6.创建一个登录 JSP 界面（有 7 种出错信息的提示），命名为 enter.jsp。登录 enter.jsp 并且不输入学号和密码，系统将会返回“请输入学号!!”提示信息，并停留在登录界面。

源代码如下：

```
<html>
    <title>课堂测试</title>
    <body>
        <%@page contentType="text/html;charset=GB2312"%>
        <%
            //让浏览器失去缓存功能
            response.setDateHeader("Expires",0);
            response.setHeader("Cache-Control","no-cache");
            response.setHeader("Pragma","no-cache");
        %>
        <center>
            <font color="blue" size="5"><i><b>课堂测试，请先输入学号和密码…</b>
              </i></font>
            <hr>
            <form action="check.jsp" method="post">
                <table border="1">
                    <tr>
                        <th bgcolor="yellow">
                            学号
                        </th>
                        <td>
                            <input type="text" size="12" name="id">
                        </td>
                    </tr>
                    <tr>
                        <th bgcolor="yellow">
                            密码
                        </th>
                        <td>
                            <input type="password" size="12" name="password">
                        </td>
                    </tr>
                    <tr>
                        <td colspan="2" align="center">
                            <input type="submit" value="测试开始">
                        </td>
                    </tr>
```

```
            </table>
        </form>
        <font color="red">
<%
    //根据错误类型,显示相应信息
    String error=request.getParameter("errortype");//由 get 返回错误参数值
    if(error!=null){
        int errortype=Integer.parseInt(error);//转换成整数
        switch(errortype){
        case 1:
            out.println("请输入学号!!");
            break;
        case 2:
            out.println("请输入密码!!");
            break;
        case 3:
            out.println("学号错误!!");
            break;
        case 4:
            out.println("密码错误!!");
            break;
        case 5:
            out.println("你参加过测试,不可重复进行!");
            break;
        case 6:
            out.println("你不能在规定时间之外进行课堂测试!");
            break;
        case 7:
            out.println("你不能在实验室之外的地方进行课堂测试!");
            break;
        default:
        }
    }
%>
    </font>
        </center>
    </body>
</html>
```

程序运行效果如图 11-9 和图 11-10 所示。

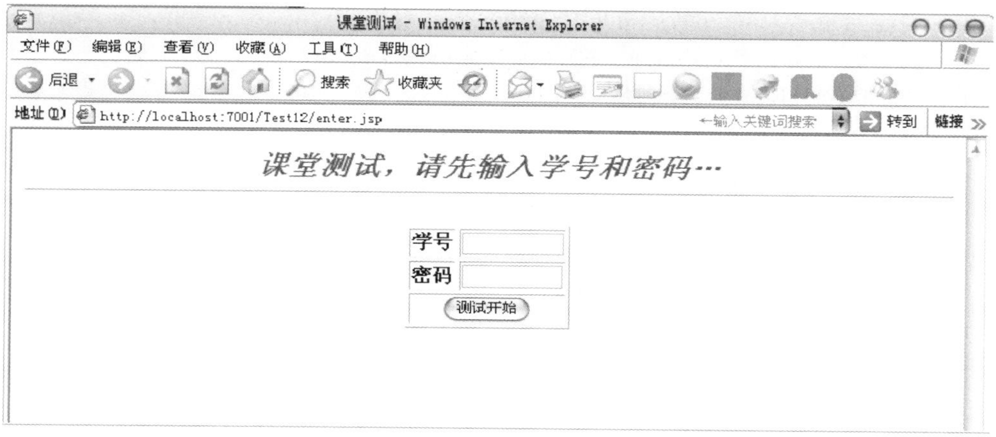

图 11-9 登录界面

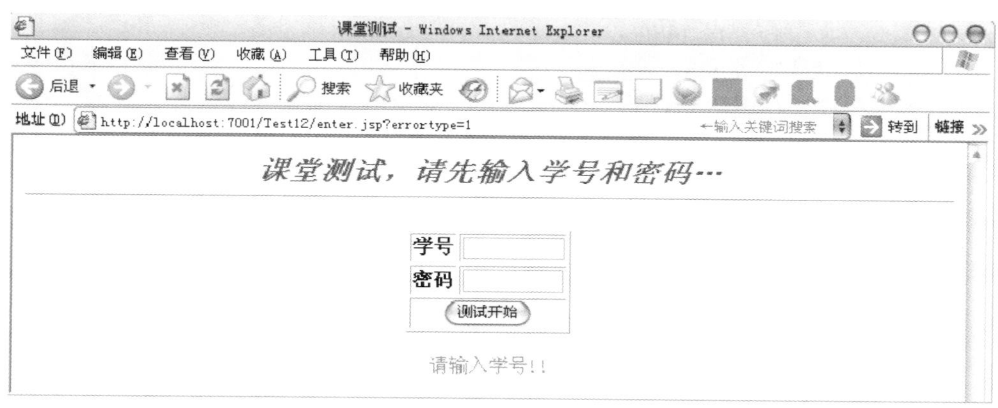

图 11-10 空输入后提示

然后建立 check.jsp 页面检验用户身份，如果学号和密码输入正确(即与 Access 数据库中的记录一致)，则跳转到 exam.jsp 页面。

当出现下列情况时将会要求再次登录，并且做相应的提示：

(1)未输入学号或者密码。

(2)学号对应学生所在班级的班号不在对应的时间段内。

(3)学号与密码不匹配。

(4)这个学号的学生已经测试过(判断测试的学生所使用的 IP 是否已经存入数据库对应的 IP 字段)。

(5)参加测试的学生访问 IP 网段未经授权(默认为 127.0.＊.＊网段)。

源代码如下：

```
<%@page contentType="text/html;charset=GB2312"%>
<%@page import="java.util.* "%>
<%@page import="java.sql.* "%>
<%
    //从 web.xml 中获取系统初始参数(指定保存的 ip)、Access 文件的路径和名字
```

```
String myip=application.getInitParameter("ip");
String accessfile=application.getInitParameter("accessfile");
//获取系统当前日期和时间值
Calendar calendar=new GregorianCalendar();
int dayOfWeek=calendar.get(Calendar.DAY_OF_WEEK);
int hourOfDay=calendar.get(Calendar.HOUR_OF_DAY);
int minute=calendar.get(Calendar.MINUTE);
String classID=null;
//根据时间给出正在上课学生的班号
if((dayOfWeek==2)&&(hourOfDay==11)&&(minute>=30)
      &&(minute<=50))
   classID="FK01";
else if((dayOfWeek==2)&&(hourOfDay==10)&&(minute>=0)
      &&(minute<=20))
   classID="FK02";
else if((dayOfWeek==5)&&(hourOfDay==18)&&(minute>=0)
      &&(minute<=20))
   classID="FK03";
else if((dayOfWeek==3)&&(hourOfDay==16)&&(minute>=30)
      &&(minute <=50))
   classID="FJ01";
else if((dayOfWeek==5)&&(hourOfDay==16)&&(minute>=30)
      && (minute<=50))
   classID="FJ02";
else if((dayOfWeek==6)&&(hourOfDay==16)&&(minute>=30)
      &&(minute<=50))
   classID="FJ03";
else
   classID="NONE";
//为测试方便使用的班级 ID
classID ="FJ01";
//获取输入的学号和密码
String id=request.getParameter("id");
String password=request.getParameter("password");
int errortype=0;//声明错误代码变量
if (id==null||id.equals("")){          //未输入学号
   errortype=1;
   response.sendRedirect("enter.jsp? errortype=" + errortype);
} else if(password==null||password.equals("")){//未输入密码
   errortype=2;
   response.sendRedirect("enter.jsp? errortype=" + errortype);
}
```

```
else{
    //连接数据库,检查是否为合法用户
    //直接用路径访问 Access 文件
    Class.forName("sun.jdbc.odbc.JdbcOdbcDriver");
    Connection con = DriverManager. getConnection (" jdbc: odbc: driver =
      {microsoft access driver (*.mdb)};dbq="+accessfile);
    Statement smt=con.createStatement();
    String sql="select* from students where id='" +id+"'";
    ResultSet rs=smt.executeQuery(sql);//检查数据库中有无指定学号的学生
    if(rs.next()){//若存在该学号的学生,则获取其密码、班号、姓名、IP 等信息
        String lib_password=rs.getString("password");
        String lib_classid=rs.getString("classid");
        //获取数据库指定 IP(如 ip1、ip2...)
        String lib_ip=rs.getString(myip);
        if(lib_ip==null)
            lib_ip="";
        String lib_name=rs.getString("name");
        if(!lib_password.equals(password)){ //密码错误
            errortype=4;
            response.sendRedirect("enter.jsp? errortype="+errortype);
        }
        else if(!lib_ip.equals("")){ //已测试过,已记录成绩
            errortype=5;
            response.sendRedirect("enter.jsp? errortype="+errortype);
        }
        else if(!lib_classid.equals(classID)){ //班号错误,即不在规定时间考试
            errortype=6;
            response.sendRedirect("enter.jsp? errortype="+ errortype);
        }
        else if (! classID. equals (" NONE") &&! request. getRemoteAddr ( ).
        startsWith("127.0.")){ //网段错误,即不在规定的网段不能登录
            errortype=7;
            response.sendRedirect("enter.jsp? errortype="+ errortype);
        }
        else{//一切正常,进入考试阶段
            //在 session 中保存学号、姓名和班号,并派发"通行证"
            session.setAttribute("id",id);
            session.setAttribute("classid",classID);
            session.setAttribute("name",lib_name);
            session.setAttribute("permitted","yes");
            //可用请求分派或重定向方式,进入考场
            //RequestDispatcher rd=request.getRequestDispatcher("exam.jsp");
```

```
            //rd.forward(request,response);
            response.sendRedirect("exam.jsp");
        }
    }
    else{ //学号错误
        errortype=3;
        response.sendRedirect("enter.jsp? errortype="+errortype);
    }
    con.close();
}
%>
```

7.学号和密码验证通过后,进入测试页面 exam.jsp。

源代码如下:

```
<%@page contentType="text/html;charset=GB2312" import="java.sql.* "%>
<%
    //强制使浏览器失去缓存功能
    response.setDateHeader("Expires",0);
    response.setHeader("Cache-Control","no-cache");
    response.setHeader("Pragma","no-cache");
    //检查是否为合法用户,如果不是,则重新登录
    String permitted=(String) session.getAttribute("permitted");
    if(permitted==null||!permitted.equals("yes"))
        response.sendRedirect("enter.jsp");
    //从 session 中得到考生学号、姓名和班号
    String id=(String)session.getAttribute("id");
    String name=(String)session.getAttribute("name");
    String classid=(String)session.getAttribute("classid");
    if(id==null||name==null||classid==null||id.equals("")||name.equals("")||
    classid.equals(""))
        response.sendRedirect("enter.jsp");
    //获取系统初始参数(Excel 文件名以及判断题、单选题、多选题每小题的分值)
    String excelfile=application.getInitParameter("excelfile");
    String value1=application.getInitParameter("value1");
    String value2=application.getInitParameter("value2");
    String value3=application.getInitParameter("value3");
%>
<html>
    <title>考试中…</title>
    <body>
```

```
<body>
    <%
        if(!name.equalsIgnoreCase("cms")&&!name.equalsIgnoreCase("lrq")){
        //教师可以复制题目,学生则不允许
    %>
<body oncontextmenu="return false" onfragstart="return false"
    onselectstart="return false" onselect="document.selection.empty()"
    oncopy="document.selection.empty()" onbeforecopy="return false"
    onmouseup="document.selection.empty()">
    <%
        }
        else{
    %>
<body>
    <%
        }
        //输出考生信息
    %>
    <h2>
        考生姓名:
        <font color="green"><%=name%></font>   学号:
        <font color="red"><%=id%></font>   班级:
        <font color="red"><%=classid%></font>
    </h2>
    <hr>
    <form action="grades.jsp" action="post">
        <%
        //直接用路径访问 Excel 文件
        Class.forName("sun.jdbc.odbc.JdbcOdbcDriver");
        Connection con=DriverManager.getConnection("jdbc:odbc:driver=
        {microsoft excel driver (*.xls)};dbq="+excelfile);
        Statement smt=con.createStatement();
        String sql;
        int i=1;//小题编号
        sql="select* from [judges$]";
        ResultSet rs=smt.executeQuery(sql);//获取"判断题"信息
        out.println("<h3>一、判断是非题(每题" + value1+ "分)</h3>");
        out.println("<table>");
        while(rs.next()){ //显示"判断题"内容
            out.println("<tr><td>" +i+". ");
            out.println(rs.getString("question")+ "</td></tr>");
            out.println("<tr><td><font color=blue><b>  
```

```
          正确<input type='radio' name='a'
        + i+"' value='T'>   错误<input type='radio' name
        ='a"+i+"'value='F'></b></font></td></tr>");
        i++;
    }
    out.println("</table>");
    out.println("<h3>二、单项选择题(每题"+value2+"分)</h3>");
    i=1;
    sql="select* from [chooses$]";//获取"单选题"信息
    rs=smt.executeQuery(sql);
    out.println("<table>");
    while(rs.next()){ //显示"单选题"内容
        out.println("<tr><td>"+i+". </td><td>");
        out.println(rs.getString("question")+ "</td></tr>");
        out.println("<tr><td><input type='radio' name='b"+i
                + "' value='A'></td><td><font color=blue><b>A."
                +rs.getString("A")+ "</b></font></td></tr>");
        out.println("<tr><td><input type='radio' name='b"+ i
                + "' value='B'></td><td><font color=blue><b>B."
                +rs.getString("B")+ "</b></font></td></tr>");
        out.println("<tr><td><input type='radio' name='b"+ i
                + "' value='C'></td><td><font color=blue><b>C."
                +rs.getString("C")+ "</b></font></td></tr>");
        out.println("<tr><td><input type='radio' name='b"+ i
                + "' value='D'></td><td><font color=blue><b>D."
                +rs.getString("D")+ "</b></font></td></tr>");
        i++;
    }
    out.println("</table>");
    out.println("<h3>三、多项选择题(每题"+value3+"分)</h3>");
    i=1;
    sql="select* from [multichooses$]";//获取"多选题"信息
    rs=smt.executeQuery(sql);
    out.println("<table>");
    while(rs.next()){ //显示"多选题"内容
        out.println("<tr><td>"+i+". </td><td>");
        out.println(rs.getString("question")+ "</td></tr>");
        out.println("<tr><td><input type='checkbox' name='c"+i
                + "' value='A'></td><td><font color=blue><b>A."
                +rs.getString("A")+ "</b></font></td></tr>");
        out.println("<tr><td><input type='checkbox' name='c"+i
```

```
                    +"' value='B'></td><td><font color=blue><b>B."
                    +rs.getString("B")+"</b></font></td></tr>");
                out.println("<tr><td><input type='checkbox' name='c"+i
                    +"' value='C'></td><td><font color=blue><b>C."
                    +rs.getString("C")+"</b></font></td></tr>");
                out.println("<tr><td><input type='checkbox' name='c"+i
                    +"' value='D'></td><td><font color=blue><b>D."
                    +rs.getString("D")+"</b></font></td></tr>");
                i++;
            }
            out.println("</table>");
            con.close();
            %>
        <hr>
        <center>
            <input type="submit" value="交卷">
        </center>
        </form>
    </body>
</html>
```

exam.jsp 页面效果如图 11-11 所示。

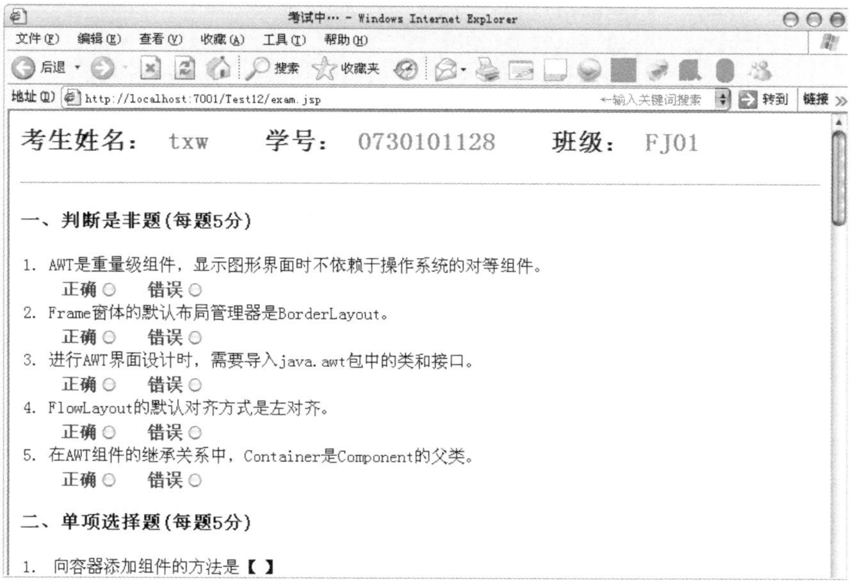

图 11-11　测试答题界面

8.最后，建立 grades.jsp 页面，根据 Excel 里面定义的答案，将考试成绩输出显示。源代码如下：

```
<%@page contentType="text/html;charset=GB2312" import="java.sql.*"%>
```

```
<%
    //让浏览器失去缓存功能
    response.setDateHeader("Expires",0);
    response.setHeader("Cache-Control","no-cache");
    response.setHeader("Pragma","no-cache");
    //从 session 中得到考生学号、姓名和班号
    String id=(String)session.getAttribute("id");
    String name=(String)session.getAttribute("name");
    //检查是否为合法用户,如果不是,则重新登录
    String permitted=(String)session.getAttribute("permitted");
    if(permitted==null||!permitted.equals("yes"))
        response.sendRedirect("enter.jsp");
    //获取系统初始参数的值(Access、Excel 文件的路径以及保存的 ip、score 和各题分值)
    String accessfile=application.getInitParameter("accessfile");
    String excelfile=application.getInitParameter("excelfile");
    String ip=application.getInitParameter("ip");
    String score=application.getInitParameter("score");
    String value1=application.getInitParameter("value1");
    String value2=application.getInitParameter("value2");
    String value3=application.getInitParameter("value3");
    //转换成整数
    int v1=Integer.parseInt(value1);
    int v2=Integer.parseInt(value2);
    int v3=Integer.parseInt(value3);
%>
<html>
    <title>保存考试成绩</title>
    <body>
        <%
        //直接用路径访问 Excel 文件
        Class.forName("sun.jdbc.odbc.JdbcOdbcDriver");
        Connection con=DriverManager.getConnection("jdbc:odbc:driver=
        {microsoft excel driver (*.xls)};dbq="+excelfile);
        Statement smt=con.createStatement();
        String sql;
        String answer,key;
        int i;
        int grades=0;
        //计算"判断题"分值
        sql="select key from [judges$]";
        ResultSet rs=smt.executeQuery(sql);
```

```
i=1;
while(rs.next()){
    key=rs.getString("key");//取得正确答案
    try{
        answer=request.getParameter("a"+i);
        if (answer.equalsIgnoreCase(key)) //比较答案是否正确,正确则加分
            grades=grades+v1;
    }catch(Exception e){
        answer="";
    }
    i++;
}
//计算"单选题"分值
sql="select key from [chooses$]";
rs=smt.executeQuery(sql);
i=1;
while(rs.next()){
    key=rs.getString("key");
    try{
        answer=request.getParameter("b"+i);
        if(answer.equalsIgnoreCase(key))
            grades=grades+v2;
    }catch(Exception e){
        answer="";
    }
    i++;
}
//计算"多选题"分值
sql="select key from [multichooses$]";
rs=smt.executeQuery(sql);
i=1;
while(rs.next()){
    key=rs.getString("key");
    try{    //获得复选框选项
        String options[]=request.getParameterValues("c"+i);
        answer="";
        if(options!=null){
            for(int j=0;j<options.length;j++)
                answer+=options[j];
        }
```

```
            if(answer.equalsIgnoreCase(key))
                grades=grades+v3;
        }catch(Exception e){
            answer="";
        }
        i++;
    }
    con.close();
    //直接连接 Access 文件,再次打开 students 数据库,保存学生成绩
    con=DriverManager.getConnection("jdbc:odbc:driver={microsoft
      access driver (*.mdb)};dbq="+accessfile);
    smt=con.createStatement();
    sql="update students set "+score+ "="+grades+ ","+ip
          +"='"+request.getRemoteAddr()+ "'where id='"+id+"'";
    System.out.println(sql);
    smt.executeUpdate(sql);
    con.close();
%>
    <center>
        <p></p>
        <p></p>
        <p></p>
        <p></p>
  <font color="blue" size="6">
  <%
    if (name==null)
        out.println("数据已保存,你的重考成绩无效!");
    else
        out.println(name+ "同学的成绩:"+grades+ "分");
     session.invalidate();//强制结束会话
%>
</font>
    <h1>
        本次课堂测试结束!
    </h1>
  </center>
  </body>
</html>
```

程序运行效果如图 11-12 所示。

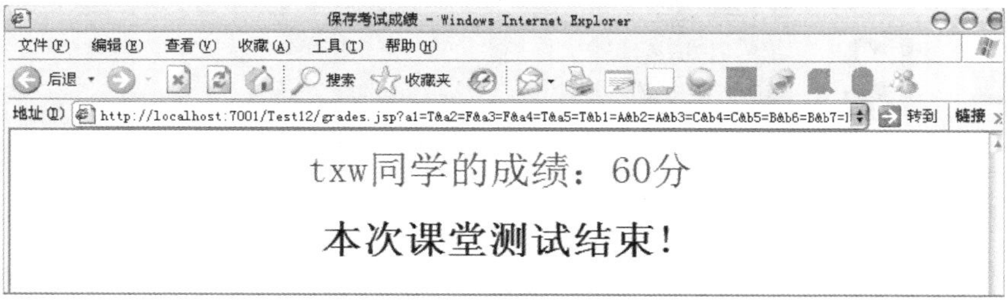

图 11-12 统计本次测试成绩效果图

11.5 实验总结

JSP 的隐式对象共有 9 个:pageContext、request、session、application、page、config、response、out 和 exception;JSP 脚本包括声明、表达式、Scriptlet(脚本)和转义字符;JSP 的注释分为 HTML 注释和 JSP 注释;JSP 定义的作用域包括 page、request、session 和 application。本章实验内容为 JSP 应用综合实例,通过应用 JSP 的元素、指令、隐式对象以及注释等,让读者体验到应用 JSP 技术可以更好地设计客户端页面,也可以将数据库的数据在客户端中方便地显示,进一步加深了对 JSP 技术的理解。

11.6 课后思考题

1.page 指令中最常用的属性有哪些?

2.include 指令有什么功能?

3.简述 JSP 隐式定义的 9 大对象的名称和主要功能。

4.JSP 中的声明、表达式和脚本有什么区别?

实验 12

JDBC 与 JSP 实践

12.1 实验目的

1.理解 JDBC 的工作原理
2.掌握获取数据库连接的代码编写技巧
3.掌握 Tomcat 下的 JDBC-JNDI 数据源库的配置方法
4.掌握在 JSP 页面中对数据进行增、删、改、查等操作
5.提高代码的阅读、分析能力

12.2 实验环境

1.MyEclipse 插件平台
2.Weblogic、Tomcat 容器
3.MySQl(或者 SQLServer、Oracle)数据库

12.3 实验知识背景

12.3.1 JDBC 基础

1.JDBC 技术简介

JDBC 是 Java DataBase Connectivity(Java 数据连接)技术的简称,是一种可用于执行 SQL 语句的 Java API。它由一些 Java 语言编写的类和接口组成。

JDBC 为数据库应用开发人员、数据库前台工具开发人员提供了一种标准的应用程序设计接口,使开发人员可以用纯 Java 语言编写完整的数据库应用程序。

JDBC 主要功能有:与数据库建立连接;向数据库发送 SQL 语句;处理数据返回的结果。

2.JDBC 工作原理

JDBC 有两个程序包：

- java.sql：核心包，这个包中的类主要完成数据库的基本操作，如生成连接、执行 SQL 语句、预处理 SQL 语句等；
- javax.sql：扩展包，主要为数据库方面的高级操作提供了接口和类。

JDBC 常用类和接口：

- Driver 接口：加载驱动程序；
- DriverManager 类：装入所需的 JDBC 驱动程序，编程时调用它的方法来创建连接；
- Connection 接口：编程时使用该类对象创建 Statement 对象；
- Statement 接口：编程时使用该类对象得到 ResultSet 对象；
- ResultSet 类：负责保存 Statement 执行后所产生的查询结果。

JDBC 的基本工作原理就是通过这些 API 来实现与数据库建立连接、执行 SQL 语句、处理结果等操作，如图 12-1 所示。

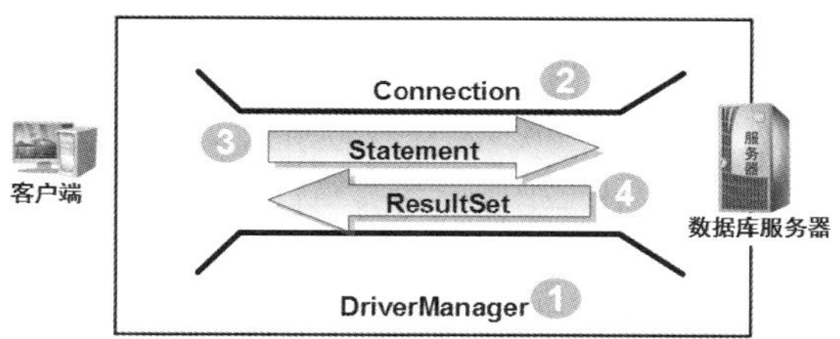

图 12-1 JDBC 工作原理示意图

12.3.2 JDBC 工作方式

JDBC 目前使用较多的工作方式为 JDBC-ODBC 桥连方式和纯 Java 驱动方式，如图 12-2 所示。

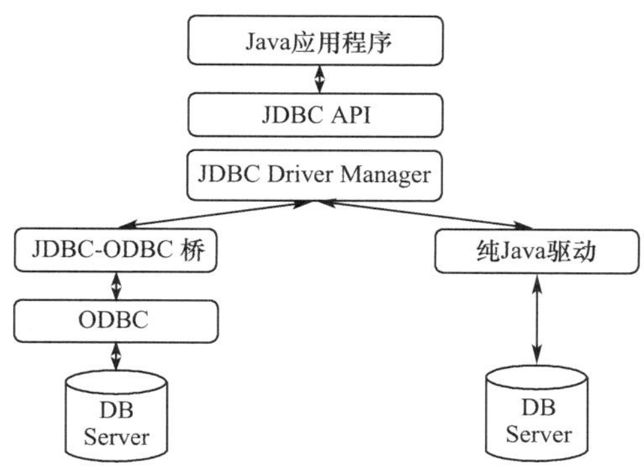

图 12-2 JDBC 桥接与驱动方式示意图

桥连方式本质上是将对 JDBC API 的调用转换为对另一组数据库连接 API 的调用。这种工作方式的优点是可以访问所有 ODBC 可以访问的数据库,缺点也比较明显,执行效率低、功能不够强大。JDBC-ODBC 桥接工作方式如图 12-3 所示。

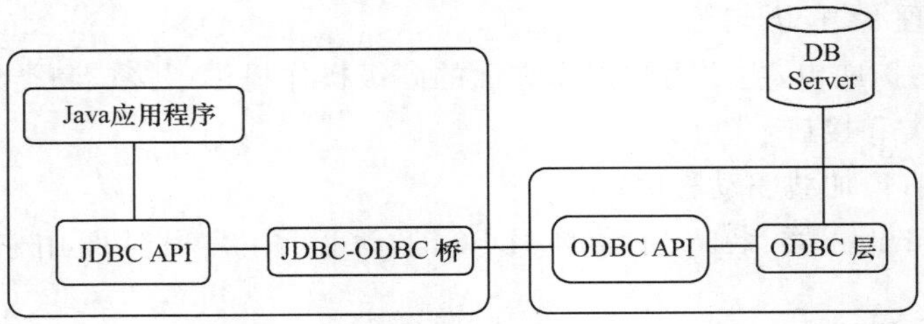

图 12-3　JDBC-ODBC 桥接工作方式

纯 Java 驱动方式是使用由数据库供应商提供的 JDBC 驱动直接访问数据库的操作方式,这种方式的优点为 100%使用 Java 语言开发,开发效率高,跨平台性能好;缺点是当访问不同的数据库需要下载专用的 JDBC 驱动。JDBC 纯 Java 驱动工作方式如图 12-4 所示。

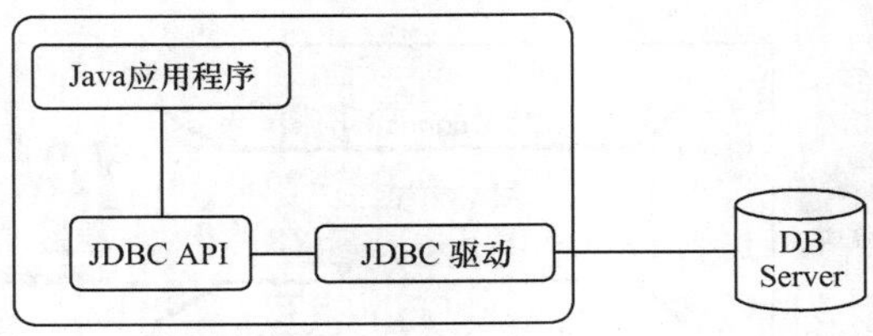

图 12-4　JDBC 纯 Java 驱动工作方式

12.3.3　JDBC 技术的使用

JDBC 的基本操作步骤:

- 创建数据源(使用 JDBC-ODBC 桥式驱动程序时必须创建);
- 注册、加载特定的驱动程序;
- 创建 Connection 连接对象;
- 利用 Connection 对象生成 Statement 对象;
- 利用 Statement 对象执行 SQL 语句,如增、删、改、查等;
- 若是执行查询语句,还要从 ResultSet 读取数据;
- 关闭 ResultSet、Statement、Connection 等。

下面是一个使用 JDBC 技术操作数据库的示例:

```
try {
    Class.forName(JDBC 驱动类);                    ①注册JDBC 驱动
} catch (ClassNotFoundException e) {
    System.out.println("无法找到驱动类");
try {                                              ②获得数据库连接
    Connection con= DriverManager.getConnection(JDBC URL,数据库用户名,密码);
                                                   ③JDBC URL 标识要访问的数据库
```

```
        Statement stmt =  con.createStatement();
        ResultSet rs =  stmt.executeQuery("SELECT a, b, c FROM Table1");
        while (rs.next()) {
                int x =  rs.getInt("a");
                String s =  rs.getString("b");
                float f =  rs.getFloat("c");
        }
        con.close();
    } catch (SQLException e) {
        e.printStackTrace();
    }
```


12.3.4 预处理语句在 JDBC 中的应用

当向数据库发送一个 SQL 语句，比如“Select * from student”，数据库中的 SQL 解释器负责把 SQL 语句生成底层的内部命令，然后执行该命令，完成有关的数据操作；如果不断地向数据库提交 SQL 语句势必增加数据库中 SQL 解释器的负担，影响执行的速度；如果应用程序能针对连接的数据库，事先就将 SQL 语句解释为数据库底层的内部命令，然后直接让数据库去执行这个命令，则不仅能减轻数据库的负担，也能提高访问数据库的速度。

针对上述情况，在 JDBC 的实际应用中，如果使用 Connection 和某个数据库建立了连接对象 conn，那么 conn 就可以调用 preparedStatement(String sql)方法对 SQL 语句进行预编译处理，生成该数据库底层的内部命令，并将该命令封装在 PreparedStatement 对象中，该对象调用相应的方法都可以使得该底层的内部命令被数据库执行。在创建 PreparedStatement 对象时，SQL 命令语句是作为参数提供的。由于 SQL 命令中的未知值的位置是已知的，使用“?”来表示，运行 SQL 语句时，可以根据需要设置实际值。

例如：

```
PreparedStatement ps;
ps =  conn.preparedStatement("insert into table (col1,col2) values (?, ?)");
ps.setInt(1,100);
ps.setString(2,"Dennis");
ps.execute();
```

12.4 实验内容与步骤

12.4.1 JDBC-ODBC 桥接方式

1.使用 JDBC-ODBC 桥接方式实现 ILIKE 信息管理系统的管理员身份登录功能，要求使用会话跟踪技术在主页面上显示管理员的身份，具体业务流程如图 12-5 所示。

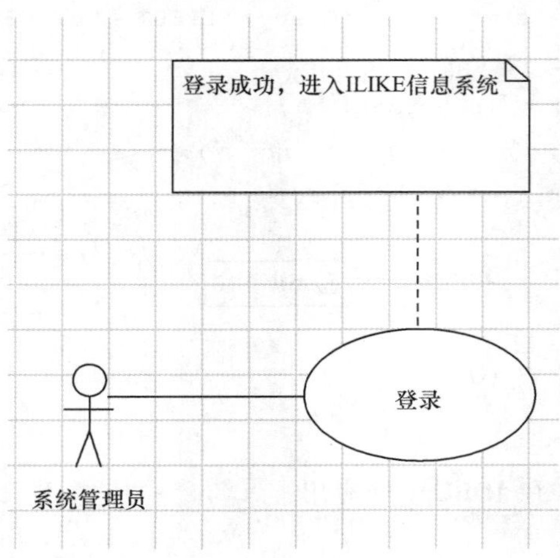

图 12-5　系统管理员登录模块例图

(1)新建 Access 数据库 db_ilike.mdb，在数据库中建立表 admin_tb，表结构如图 12-6 所示。建好表后在表中输入一条记录，作为登录测试使用。

字段名称	数据类型
adminID	文本
adminPswd	文本
adminName	文本

图 12-6　admin_tb 表结构

(2)打开[控制面板]|[管理工具]|[数据源(ODBC)]配置 ODBC 数据源，将建好的 Access 数据库作为数据源，命名为 db_ilike，如图 12-7 所示。

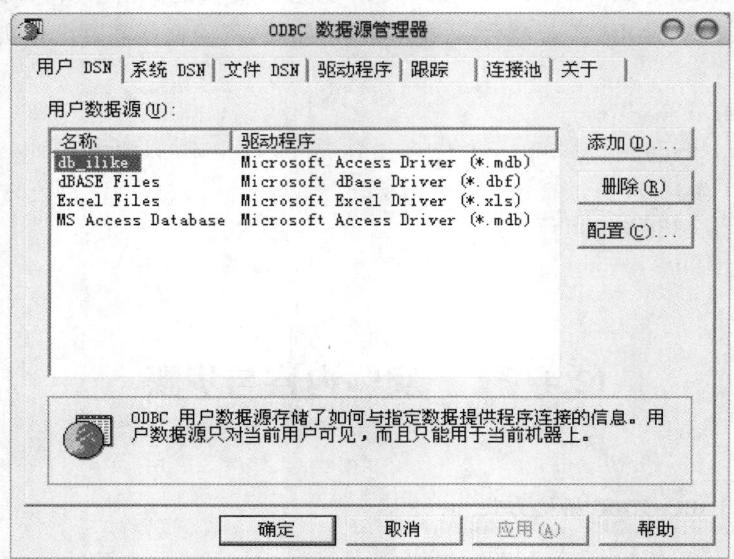

图 12-7　配置 ODBC 数据源

(3)在 MyEclipse 中新建一个 Web 项目,在项目中新建 login.jsp 页面,该页面主要功能是提供登录输入,当输入密码不正确的时候会出现提示信息。

源代码如下:

```
<%@ page contentType="text/html; charset=gb2312" %>
<html>
<head>
<title>登录</title>
<style type="text/css">
<!--
.STYLE1 {
color: #FFCC33;
font-size: 14px;
}
-->
</style>
<link href="Css/style.css" rel="stylesheet" type="text/css">
<link href="Css/styleleft.css" rel="stylesheet" type="text/css">
<meta http-equiv="Content-Type" content="text/html; charset=gb2312">
<style type="text/css">
<!--
body {
margin-top: 200px;
}
-->
</style>
</head>
<body>
<table width="500" border="0" align="center"  cellpadding="0" cellspacing="0" bgcolor="#F5f7f7">
<form name="form" method="post" action="loginCheck.jsp">
  <tr align="center">
    <td height="27" colspan="2" bgcolor="#21536A" class="STYLE1"><strong><font size="+3"> 员工登录</font></strong>
  </td>
  </tr>
  <tr>
    <td width="200" height="22" align="center" bgcolor="#F5F7F7">用 户 名:</td>
    <td width="300" bgcolor="#F5F7F7">
      <input  type="text" name="adminID">
  </td>
```

```
    < /tr>
    < tr>
      < td height= "22" align= "center" bgcolor= "# F5F7F7"> 密    
 码:< /td>
      < td bgcolor= "# F5F7F7"> < input type= "password" name= "adminPswd">
    < /td>
    < /tr>
    < tr align= "center">
      < td height= "35" colspan= "2" bgcolor= "# F5F7F7">
      < input type= "submit" name= "Submit" value= "登录">
          
        < input type= "reset" name= "Reset" value= "重置">
      < /td>
      < /tr>
        < tr>
        < td height= "22" align= "center" bgcolor= "# F5F7F7"> < /td>
        < td bgcolor= "# F5F7F7"> < font color= "red" bold>
        < ! - - //验证反馈信息 - - >
        < % = request.getParameter("message") % >
        < /font>
      < /td>
      < /tr>
    < /form>
  < /table>
  < /body>
  < /html>
```

页面效果如图 12-8 所示。

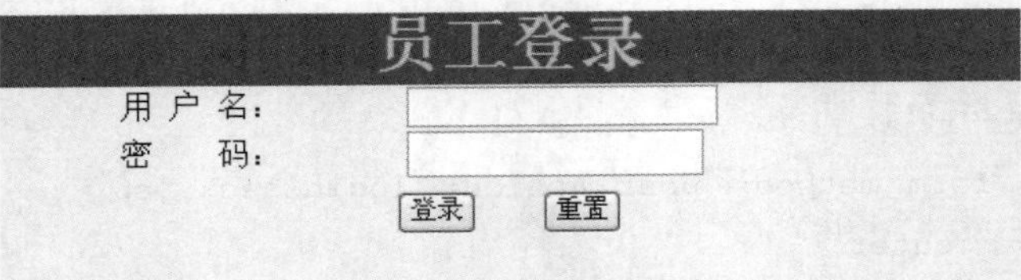

图 12-8 login.jsp 页面效果图

如果输入错误的用户名和密码,会进行提示,如图 12-9。

图 12-9 错误提示界面

(4)在项目中新建 loginCheck.jsp 页面，该页面主要功能为使用 JDBC-ODBC 桥接的方式查询 Access 数据库 db_ilike.mdb 里面的表 admin_tb，实现对用户名与密码的验证，如果验证通过则跳转至主界面，否则返回 login.jsp，并且向 login.jsp 发送出错信息。

源代码如下：

```
<%@page contentType="text/html;charset=GB2312" import="java.sql.*"%>
<% String adminID=request.getParameter("adminID");     //获取输入帐号
String adminPswd=request.getParameter("adminPswd");    //获取输入密码
String sql ="select adminPswd, adminName from admin_tb where adminID=\'"+
  adminID+"\'";    //编写 SQL 语句，根据管理员 ID，获得密码与管理员姓名
try{
    Class.forName("sun.jdbc.odbc.JdbcOdbcDriver");
    Connection con=DriverManager.getConnection("jdbc:odbc:db_ilike");
    Statement dbc=con.createStatement();
    ResultSet rs=dbc.executeQuery(sql);
    if(rs.next()){                  //判断帐号是否存在
        if(adminPswd.equals(rs.getString(1))){
        //判断用户输入密码，与数据库检索密码帐号是否一致
            session.setAttribute("adminName",rs.getString(2));//存储管理员姓名
            response.sendRedirect("index.jsp");//跳转至 index.jsp 页面
        }
    }
    else{//跳转至 login.jsp 页面
    response.sendRedirect("login.jsp? message=input data error!");
}
    rs.close();
    dbc.close();
    con.close();
    }catch(SQLException e){
        e.printStackTrace();
    }
%>
```

(5)在项目中新建 index.jsp 作为 ILIKE 系统的主界面，使用 JSP 小脚本实现阻止未经验证的用户直接访问主界面的情况出现。

源代码如下：

```
<%@page contentType="text/html;charset=GB2312"%>
<%
if(session.getAttribute("adminName")==null){
//判断是否为合法访问
    response.sendRedirect("login.jsp? message=input data error!");
//若为非法访问，跳转至 login.jsp 页面
}
```

```
%>
<html>
<head>
<title>ILIKE</title>
</head>
<frameset name="head_main" rows="60,18,* " cols="* " frameborder="no" border=
"0" framespacing="0">
  <frame src="top.html" name="top" scrolling="No" noresize="noresize" id=
  "topFrame" title="topFrame" />
< frame border = 0 name = " head _ bar" marginWidth = 0 borderColor = # e7e7e7
marginHeight=0 src="head_bar.jsp" frameBorder=0 noResize scrolling=no>
<frameset name="menu_main" cols=200,8,*  bordercolor="#3164FE">
    <frame src="left.html" name="left" scrolling="yes" noresize="noresize" id="
    leftFrame" title="leftFrame" />
    <frame border=0 name="bar" marginWidth=0 borderColor=#e7e7e7
    marginHeight=0 src="left_bar.htm" frameBorder=0 noResize scrolling=no>
    <frame src="mainMessage.html" name="main" id="mainFrame" title="" />
</frameset>
</frameset>
<noframes>
<body>
</body>
</noframes>
</html>
```

(6)在项目中新建 head_bar.jsp 页面,该页面主要显示系统的标题信息,如登录人的身份、系统时间显示等,同样要阻止未经验证的用户直接访问页面。

源代码如下:

```
<%@page contentType="text/html;charset=GB2312" %>
<%
 if(session.getAttribute("adminName")==null){
   response.sendRedirect("login.jsp? message=error");
 }
 % >
<html>
<head>
    <title>head</title>
<link href="css/stylemain.css" rel="stylesheet" type="text/css" />
    <style>
       .jd{
          font-family: Webdings;
          color: # FFFFFF;
```

```
            cursor: hand;
        }
    </style>
    <script language="JavaScript" type="text/javascript">
        var timeid=null;
        function showtime(){
            var now=new Date();
            var hours=now.getHours();
            var minutes=now.getMinutes();
            var seconds=now.getSeconds();
            timevalue=""+((hours>12)? hours-12:hours);
            timevalue+=((minutes<10)?":0":":")+minutes;
            timevalue+=((seconds<10)?":0":":")+seconds;
            timevalue=((hours>12)?"下午:":"上午:")+timevalue;
            mytimer.innerHTML="<font color=#FFFFFF><nobr> "+
            timevalue+" </nobr></font>";
            timeid=setTimeout("showtime()",1000);
        }
    </script>
</head>
<body onLoad="showtime()" bgcolor="#669999">
<a title="全屏/恢复">
<table border=0 cellpadding=0 cellspacing=0 width="100%">
<tr>
    <td width="1%" onClick="return FullScr();">
<div id="h" class="jd">1</div></td>
    <td width="97%" align="left" onClick="return FullScr();">
    <font color="white">
    <!--//显示管理员姓名  -->
     <%=session.getAttribute("adminName")%>欢迎您
    </font>
    </td>
    <td width="1%" align="right">
    <img src="images/top/clock.gif" width="11" height="11"
    border="0">
    </td>
    <td width="1%" align="right"><div id="mytimer"></div>
    </td>
</tr>
</table>
</a>
<SCRIPT language=JavaScript type=text/javascript>
```

```
    function FullScr(){
        var hc=h.innerHTML;
        if(hc=='1'){
            parent.head_main.rows='0,18,*';
            h.innerHTML='2';
        }
        else{
            parent.head_main.rows='60,18,*';
            h.innerHTML='1';
        }
        return false;
    }
</SCRIPT>
</body>
</html>
```

(7)整个 Web 项目设计完成后，部署到 WebLogic 上发布，整体效果如图 12-10 和图 12-11 所示。

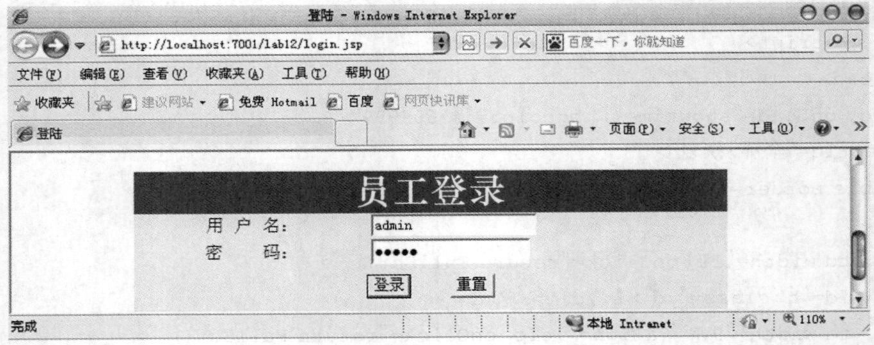

图 12-10 输入正确的用户名和密码

图 12-11 通过验证后显示主界面并识别登录用户身份

12.4.2 Tomcat 下 JDBC-JNDI 数据源方式

1.使用 Tomcat 容器配置创建 JDBC-JNDI 数据源的方式访问 MySQL 数据库,实现查询显示制造商基本资料的功能。具体步骤如下:

(1)下载 Tomcat(这里用 tomcat8.0),解压运行安装,如图 12-12 所示。

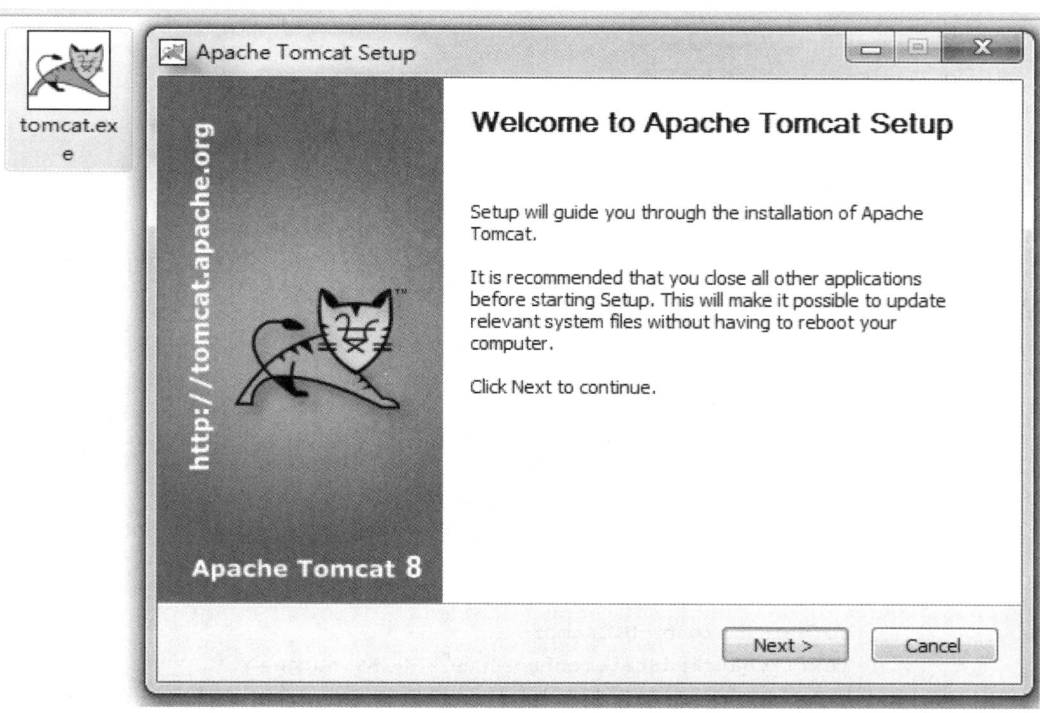

图 12-12　Tomcat 安装示意图

(2)安装过程中弹出的"Tomcat Administrator Login"界面,如图 12-13 所示,用户名和密码都可以不设置。

Server Shutdown Port　-1
HTTP/1.1 Connector Port　8080
Windows Service Name　Tomcat8
Create shortcuts for all users
Tomcat Administrator Login (optional)　User Name　tomcat
Password　••••••••
Roles　admin-gui,manager-gui

图 12-13　"Tomcat Administrator Login"界面

(3)数据库以 MySQL(这里用 MySQL8.0)为例,安装好 Tomcat 容器后,需要将 MySQL jar 包导入 Tomcat 安装目录的 lib 目录下,如图 12-14 所示。

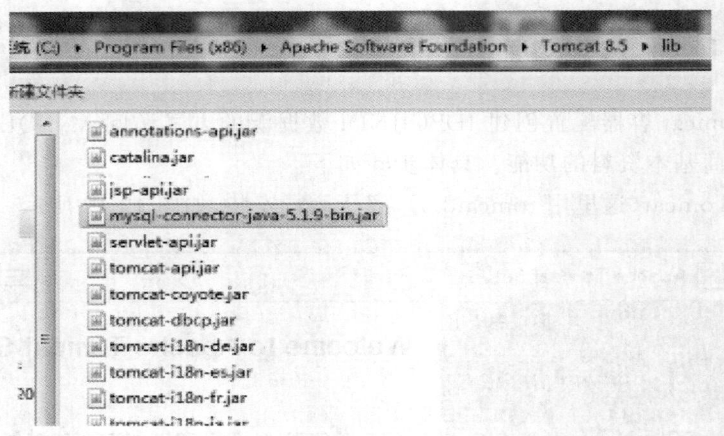

图 12-14　导入 MySQL jar 包

(4)在 Tomcat 安装目录下打开 conf 目录,打开 context.xml 文件,在里面添加设置 JDBC-JNDI 数据源,使用 Resource 标签。

```
< Resource name = "jdbc/数据库名"(测试项目 web.xml 中的设置要与此一致)
        auth = "Container"
        driverClassName= "com.mysql.cj.jdbc.Driver"
        type= "javax.sql.DataSource"(固定)
        username = "username"(数据库用户名)
        password = "password"(数据库密码)
        url = "jdbc:mysql://localhost:3306/(数据库名)
        characterEncoding= utf8&amp;
        useSSL= false&amp;
        serverTimezone= UTC&amp;
        rewriteBatchedStatements= true"> < /Resource>
```

MySQL6.0 及之后版本需要指定服务器时区属性,设定 useSSL 属性等,各属性之间用 & 连接,在 xml/html 文件中 & 用 &转义表示。

实际测试例子中的设置如图 12-15 所示。

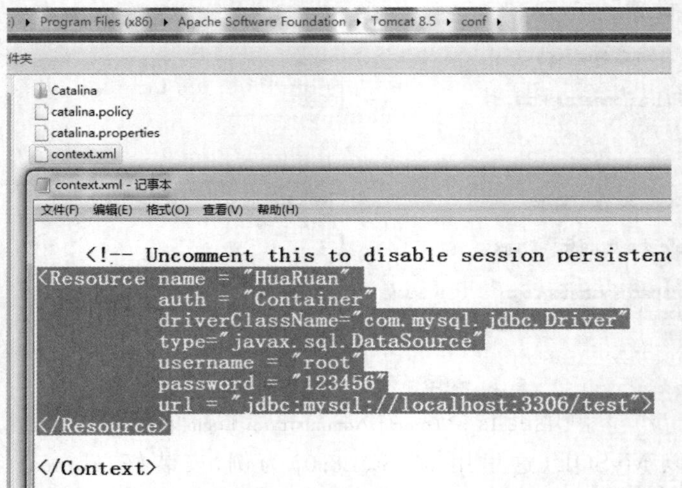

图 12-15　context.xml 中设置 JDBC-JNDI 数据源

(5)数据库表结构如图 12-16 所示。

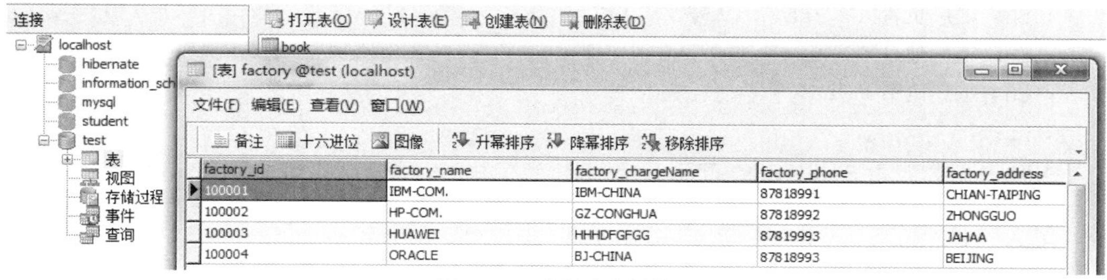

factory_id	factory_name	factory_chargeName	factory_phone	factory_address
100001	IBM-COM.	IBM-CHINA	87818991	CHIAN-TAIPING
100002	HP-COM.	GZ-CONGHUA	87818992	ZHONGGUO
100003	HUAWEI	HHHDFGFGG	87819993	JAHAA
100004	ORACLE	BJ-CHINA	87818993	BEIJING

图 12-16 数据库表结构

(6)MyEclipse 开发环境中也需要将 Tomcat 整合进来，整合方法和步骤与插入 Weblogic 服务器方法相当，参见实验 1 相应内容。

(7)在 MyEclipse(或者 Idea)下创建 Web Project，项目结构如图 12-17 所示。

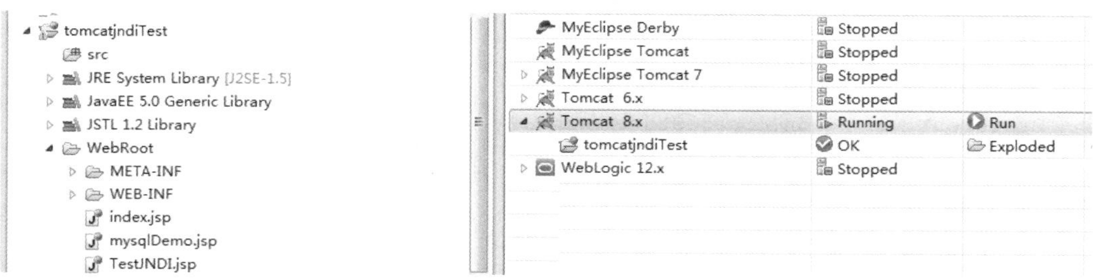

图 12-17 项目结构示意图

在 Web 项目中打开 Web.xml，编辑修改文件，添加<resource-ref></resource-ref>标签中声明对 JDBC 连接池(JNDI)的引用，如图 12-18 所示修改声明文本。

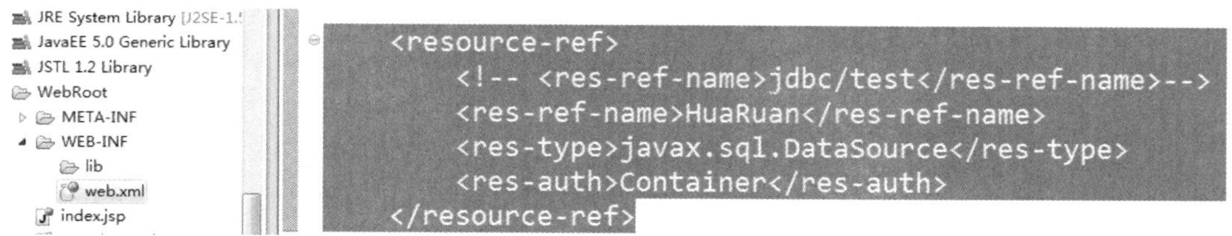

图 12-18 修改声明文本

(8)新建 JSP 页面，测试数据库连接和数据显示结果，测试 JSP 代码如下。

```
<%@ page contentType="text/html;charset=GB2312"%>
<%@ page import="javax.naming.Context"%>
<%@ page import="javax.naming.InitialContext"%>
<%@ page import="javax.sql.DataSource"%>
<%@ page import="java.sql.Connection"%>
<%@ page import="java.sql.Statement"%>
<%@ page import="java.sql.ResultSet"%>
<HTML>
<BODY>
<table border="1">
<tr>
<th>工厂编号</th>
```

```
< th> 工厂名称< /th>
< th> 工厂负责人< /th>
< th> 工厂地址< /th>
< th> 工厂电话< /th>
< /tr>
< %
try {
Context initCtx;
initCtx = new InitialContext();
Context envCtx = (Context) initCtx.lookup("java: comp/env");//固定,调用的是
tomcat 的环境
DataSource ds = (DataSource) envCtx.lookup("HuaRuan");
Connection con = ds.getConnection();
if (con ! = null) {
System.out.println("JNDI 方式连接成功");
Statement sta = con.createStatement();
String sql = "select factory_id, factory_name, factory_chargeName, factory_
address, factory_phone from factory";
ResultSetrSet = sta.executeQuery(sql);
while (rSet.next()) {
% >
< tr>
< td> < % = rSet.getLong(1)% > < /td>
< td> < % = rSet.getString(2)% > < /td>
< td> < % = rSet.getString(3)% > < /td>
< td> < % = rSet.getString(4)% > < /td>
< td> < % = rSet.getString(5)% > < /td>
< /tr>
< %
}
//System.out.println(rSet.getString("id"));
System.out.println("数据库数据访问成功!");

} else {
System.out.println("JNDI 方式连接失败");
}
//rs.close();
//sql.close();
} catch (Exception e1) {
out.println(e1.toString());
}
% >
```

```
< /table>
< /BODY>
< /HTML>
```

将项目部署到 Tomcat 中，正确启动 Tomcat、服务器，用浏览器测试 JSP，如果 MyEclipse，内置浏览器不能测试时可以换其他浏览器测试（这里用 Google 浏览器）。

将项目部署到 Tomcat 中，正确启动 Tomcat、服务器，用浏览器测试 JSP，如果 MyEclipse 内置浏览器不能测试可以换其他浏览器测试（这里用 Google 浏览器）。

12.5 实验总结

通过本次实验，让读者熟悉了 JDBC 的基础知识，其中包括 JDBC 的概念、功能、实现原理、驱动程序分类以及 JDBC 的主要类与接口。让读者练习了 JDBC 的主要功能的实现，包括与数据库建立连接、向数据库发送 SQL 语句和处理数据返回的结果等，着重训练了如何使用 JSP 技术实现 JDBC-ODBC 桥接方式与 JDBC 纯 Java 驱动方式来操作数据库。

12.6 课后思考题

1. 用 JDBC 连接数据库，驱动程序有哪四种？
2. 怎样创建对应于 books.xls 的数据源？
3. 用 JDBC 连接数据库有哪些主要步骤？
4. 说明下列类或接口的功能：Class、DriverManager、Connection、Statement、ResultSet。
5. 比较 JDBC 与 JNDI 获取数据源的方式的异同。

实验 13

JavaBean 和 JSP 标准操作

13.1 实验目的

1.理解 JavaBean 的功能,熟悉其结构、存放位置和实例创建的步骤
2.掌握使用 JavaBean 的两种基本方法:对象法和标签法
3.理解创建 JavaBean 实例时不同范围值的差异
4.掌握 JSP 标准操作,熟悉对应标签的使用

13.2 实验环境

1.MyEclipse 插件平台
2.WebLogic(或 Tomcat)容器

13.3 实验知识背景

13.3.1 在 JSP 中使用 JavaBean

要在 JSP 中使用 JavaBean,应把 JavaBean 置于项目的包里,JavaBean 不支持在默认包中创建和使用。当使用 MyEclipse 工具开发包含 JavaBean 的 Web 应用程序时,JavaBean 源程序通常放在 src 目录中,编译后形成的 class 文件通常位于 WEB-INF/classes 对应包的子目录中。

在 JSP 中一般使用对象法和标签法来操作 JavaBean。对象法与一般 Java 程序类似,在脚本中创建 JavaBean 对象,并调用相应的方法。这种方法通俗易懂,但代码较长。假设有一个命名为 BookBean 的 JavaBean 类,可以使用下面的 JSP 小脚本代码来进行

操作：

```
<%
    BookBean book1=new BookBean();
    book1.setBookName("企业级 Java Web 编程技术——Servlet & JSP ");
    book1.setAuthors("张屹 等");
    book1.setPrice(39.0);
    session.setAttribute("book1",book1);
%>
```

而标签法是通过一组 JSP 标准动作标签对 JavaBean 进行操作，这种操作代码简洁，推荐读者使用。对上面例子的 BookBean，可以在 JSP 中使用下面的语句进行操作：

```
<jsp:useBean id="book2" scope="session" class="mybean.BookBean"/>
<jsp:setProperty name="book2" property="name" value="企业级 Java Web 编程技
  术——Servlet & JSP" />
<jsp:setProperty name="book2" property="price" value="39.0" />
```

13.3.2 使用 JSP 标准动作操作 JavaBean

JSP 页面中可以使用 JSP 标准动作来与实现业务逻辑的 JavaBean 进行交互，JSP 标准动作在浏览器请求 JSP 页面时执行，JSP 标准动作还可以作为嵌入文件或其他页面的内容，如图 13-1 所示。

图 13-1　JSP 标准动作应用示意图

在 JSP 页面被翻译成 Servlet 源代码的过程中，当容器遇到标准动作元素时，则调用与之相对应的 Servlet 类方法来代替它，所有标准动作元素的前面都有一个 JSP 前缀作为标记，一般形式如下：

<jsp:标记名…属性…/>

有些标准动作中间还包含有标签体，即一个标准动作元素中又包含了其他标准动作元素或者其他内容，包括个体的标准动作的使用格式如下：

<jsp:标记名…属性…>
　　<jsp:标记名…属性以及参数…/>
</jsp:标记名>

标准动作使用<jsp>前缀，而且标准动作中的属性区分大小写，使用时需要注意。JSP 中规定了三个标签来操作 JavaBean，即：<jsp:useBean>、<jsp:setProperty>和<jsp:getProperty>，下面对这三个标签做简要介绍。

1.<jsp:useBean>标准动作

<jsp:useBean>的主要功能是创建或查找 Bean 对象，使用格式如下：

格式 1：

```
<jsp:useBean id="beanName"scope="page|request|session|application" class=
```

```
"package.BeanClass" type="TypeName" |beanName="BeanName" type="BeanName"|type=
"TypeName"/>
```

格式 2:

```
<jsp:useBean…>
      <jsp:setProperty…>
      <jsp:getProperty…>
</jsp:useBean>
```

各项参数含义:

id 指的是对象实例名称;scope 指的是 Bean 的作用范围,有 4 个可取值:page、request、session、application,page 仅涵盖使用 JavaBean 的页面,request 有效范围仅限于使用 JavaBean 的请求,session 有效范围在用户整个连接过程中(整个会话存在阶段均有效),application 有效范围涵盖整个应用程序,也就是对 Web 项目均有效;class 是 Bean 类的名称;beanName 指的是 Bean 的名称(不是实例对象的名称,而是 JavaBean 规范的名称,作为 java.beans.Beans 类的 instantiate()方法的参数);type 是 Bean 实例的类型,可以是本类,也可以是其父类,或其实现的接口,默认为本类。例如,有一个 JavaBean 的结构如下:MyFriend(属性:name、mobile),可以使用如下的代码对这个 JavaBean 进行属性赋值操作:

```
<jsp:useBean id="zhang" scope="page" class="sise.MyFriend"/>
<jsp:setProperty name="zhang" property="name" value="张三"/>
```

jsp:useBean>标准动作还可以包含个体,如<jsp:setProperty>动作,在第一次创建该实例的时候,会使用<jsp:setProperty>动作来进行参数的赋值操作,但如果该页面中已存在 JavaBean 类的实例,则不会再执行<jsp:setProperty>动作所进行的参数赋值操作。<jsp:useBean>标准动作也并不意味着每次都要创建一个实例,如果页面中已经存在该 JavaBean 的实例,则直接使用该实例。

2.<jsp:setProperty>标准动作

使用<jsp:setProperty>动作可以修改 JavaBean 实例中的属性变量,有两种使用形式:

(1)<jsp:setProperty>标准动作嵌入在<jsp:useBean>标准动作体内,但此时只能在 JavaBean 被创建的实例中执行,使用形式如下:

```
<jsp:useBean id="myName" …>
    <jsp:setProperty name="myName" property="someProperty"…/>
</jsp:useBean>
```

(2)在<jsp:useBean>动作的后面使用<jsp:setProperty>标准动作元素,使用形式如下:

```
<jsp:useBean id="myName" …/>
<jsp:setProperty name="myName" property="someProperty"…/>
```

代码说明:不管指定的 JavaBean 是新创建的还是直接使用的实例,在<jsp:useBean>动作之后使用的<jsp:setProperty>标准动作都会被执行。其中<jsp:setProperty>共有 4 个属性可以选择设置：

①name：必须设置，通过这个属性能唯一确定针对哪个 JavaBean 实例的属性进行设置。

②property：必须设置，该属性告诉容器需要对 JavaBean 实例中的哪些属性进行设置，这里有个特殊的用法，即把 property 属性设置为"*"，表示所有名字与 JavaBean 属性名字匹配的请求参数都被传递给相应属性的 set 方法。

③value：这个属性是可选的，指定 JavaBean 属性的值。

④param：这个属性和 value 属性不能同时使用，二者只能使用一个，当两个属性都没有在<jsp:setProperty>动作中指定时，指定的属性变量将使用 JavaBean 中的默认值(如类构造方法中的默认值)，如果使用 param 属性，容器就会把 property 指定的属性变量设置为 param 指定的请求参数的值。通过指定 param 选项，可以建立 Bean 属性与 request 参数的关联关系，当属性名与参数名不一致时，要指明属性与哪一个参数关联。当 property 属性设置为"*"时，所有同名参数与属性间可以进行赋值，且可自动转换。

3.<jsp:getProperty >标准动作

该标准动作与前一个<jsp:setProperty>标准动作相对应，用来提取指定的 JavaBean 属性值，然后转换成字符串输出，格式如下：

```
<jsp:getProperty name="beanName" property="propertyName"/>
```

该动作有两个必须要设置的属性，name：表示 JavaBean 在 JSP 中的标记；property：表示提取 JavaBean 中哪个属性的值。下面的例子是对命名为"zhang"的 JavaBean 对象的 Mobile 属性进行赋值。

```
<jsp:getProperty name="zhang" property="name" /></font><br>
手机号码：<font color=blue><%=zhang.getMobile()%></font>
```

13.3.3 其他常用 JSP 标准动作

除了操作 JavaBean 的几个标准动作外，<jsp:forward>、<jsp:include>、<jsp:param>和<jsp:plugin>也是比较常用的。

1.<jsp:forward>标准动作

<jsp:forward>标准动作允许从该标签处停止当前页面的执行，从而转向执行 page 属性指定的 JSP 页面。

格式：

```
<jsp:forward page="要转向的页面" />
```

或

```
<jsp:forward page="要转向的页面" >
       param 子标记
</jsp:forward>
```

在使用时要注意 forward 动作和 response.sendRedirect()的区别，<jsp:forward>使用同一个 request，而 response.sendRedirect()使用不同的 request。

如图 13-2 所示，假设 testForward.jsp 使用<jsp:forward>访问 Forward1.jsp，当地址栏输入 http://localhost:7001/JspDemo/chap13/testForward.jsp 时，地址栏显示不变；而 testsendRedirect.jsp 使用 response.sendRedirect()来重定向到 Forward1.jsp，当地址栏输入 http://localhost:7001/JspDemo/chap13/testRedirect.jsp 时，地址栏显示 http://localhost:7001/JspDemo/chap13/Forward1.jsp。

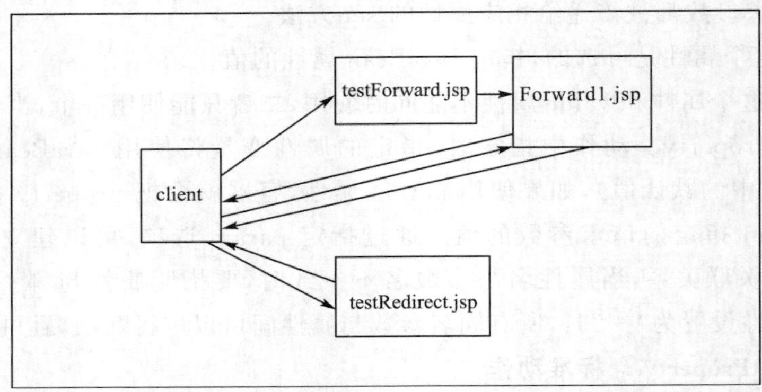

图 13-2 forward 动作和 response.sendRedirect()的区别

2.<jsp:include>标准动作

该标签的作用是使当前 JSP 页面动态包含一个文件，即将当前 JSP 页面、被包含的文件各自独立转译和编译为字节码文件。当前 JSP 页面执行到该标签处时，才加载执行被包含文件的字节码。

格式：

```
<jsp:include page="文件的 URL"/>
```

或

```
<jsp:include page="文件的 URL">
  <jsp:param name="变量名字" value="变量值" />
</jsp:include>
```

这个标签使用的时候要注意和 include 指令进行区别，<%@include file=" "%>转译时与该页面一起翻译成 Java 代码；而<jsp:include page=" "/>动作是在请求时装入该 page 所指页面。

3.<jsp:param>标准动作

<jsp:param>可以指定某个参数的值，必须和<jsp:forward>、<jsp:include>、<jsp:plugin>等协同使用。

格式：

```
<jsp:param name="attName" value="AttValue"/>
```

4.<jsp:plugin>标准动作

<jsp:plugin>元素用于在浏览器中播放或显示一个对象(典型的就是 Applet 和 Bean)。当 JSP 网页被编译后送往浏览器执行时，<jsp:plugin>会根据浏览器的版本替换成<object>或者<embed>元素。一般来说，<jsp:plugin>元素会指定对象是 Applet 还是 Bean，同样也会指定类的名字和位置，另外还会指定将从哪里下载这个 Java

组件。

语法：

```
<jsp:plugin type="bean|applet" code="objectCode" codebase="objectCodebase"
[align="alignment"][archive="archiveList"][height="height"][hspace="hspace"]
[jreversion="jreversion" ][vspace="vspace" ][width="width"][nspluginurl="URL"]
[iepluginurl="URL"]>
[<jsp:params>
 [<jsp:param value="{PV | <% =expression % >}" />]+
</jsp:params>]
 [<jsp:fallback>text message for user </jsp:fallback>]
</jsp:plugin>
```

<jsp:plugin>的参数较多，以下是对各参数的说明：

(1)type="bean|applet"：对于将被执行的对象类型必须指定是 Bean 还是 Applet，因为该属性没有默认值。

(2)code="objectCode"：即将被 Java Plugin 执行的 Java 类名称，必须以".class"结尾，并且.class 类文件必须存在于 codebase 属性所指定的目录中。

(3)codebase="objectCodebase"：如果没有设定将被执行的 Java 类的目录(或者是路径)的属性，默认使用<jsp:plugin>的 JSP 网页所在的目录。

(4)align="alignment"：图形、对象和 Applet 的位置。align 的值可以为 bottom、top、middle、left 或 right。

(5)archive="archiveList"：一些由逗号分开的路径名，用于预先加载一些将要使用的类，此做法可以提高 Applet 的性能。

(6)height="height" width="width"：设置 Applet 或 Bean 的长、宽的值，单位为像素(pixel)。

(7)hspace="hspace" vspace="vspace"：表示 Applet 或 Bean 显示时在屏幕左右、上下所需留下的空间，单位为像素 (pixel)。

(8)jreversion ="jreversion"：表示 Applet 或 Bean 执行时所需的 Java Runtime Environment (JRE)版本，默认值是 1.1。

(9)nspluginurl="URL"：表示 NetScape Navigator 用户能够使用的 JRE 的下载地址，此值为一个标准的 URL。

(10)iepluginurl="URL"：表示 IE 用户能够使用的 JRE 的下载地址，此值为一个标准的 URL。

另外，可以使用<jsp：params>传送参数给 Applet 或 Bean。也可以使用<jsp:fallback>在不能启动 Applet 或 Bean 时，在浏览器中出现一段错误提示信息。

以下是一个调用名为"Hello"的 Applet 的例子：

```
<jsp:plugin type="applet" code="test.Hello.class" codebase="."jreversion="1.5"
width="100" height="160" >
     <jsp:fallback>
     </jsp:fallback>
```

```
</jsp:plugin>
```

要注意的是，这个例子中应把编译后的 Applet 对应的 class 放在 Web 项目的根目录的/test 目录下，否则会访问失败，因为 Web 项目部署在 Web 容器后，class 文件一般是放在 Web 项目目录的 WEB-INF 目录内，JSP 页面没有访问权限，将导致访问异常。

13.4 实验内容与步骤

按要求编写 JSP 程序，实现如图 13-3 所示的"学生信息注册"模块。

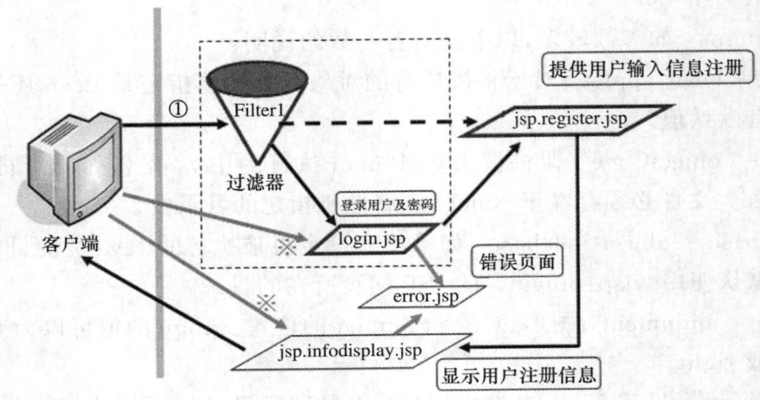

图 13-3 "学生信息注册"模块示意图

功能需求如下：

(1)客户端访问页面 jsp_register.jsp 进行信息注册时，经过过滤器 Filter1，此时把请求分派到登录页面 login.jsp，输入用户名及密码，当单击【提交】按钮后，才可以进入页面 jsp_register.jsp。

(2)当学生注册完毕后，单击【提交】按钮，进入页面 jsp_infodisplay.jsp 显示已注册信息。

(3)只有当用户访问页面 jsp_register.jsp 时，才要求经过过滤器，其他页面不经过过滤器。

若有用户想直接访问 login.jsp 页面，或想直接访问 jsp_infodisplay.jsp页面时，则属于非法进入，直接跳转至错误页面 error.jsp。

1.新建 Web 项目，项目结构如图 13-4 所示。

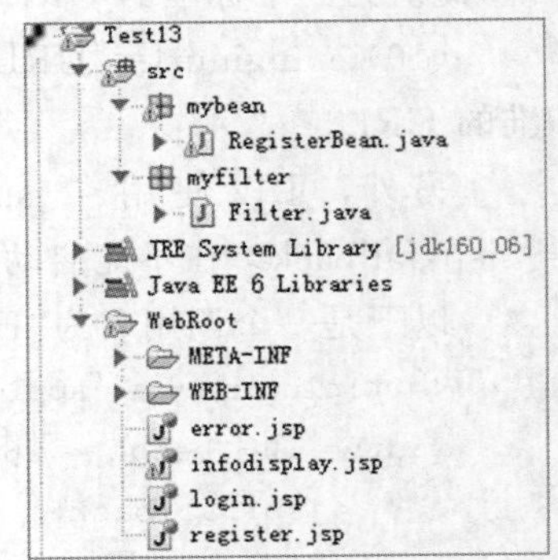

图 13-4 项目结构图

2.在项目中新建 JavaBean 类，命名为 RegisterBean。

源代码如下：

```
package mybean;
import java.io.Serializable;
import java.util.*;
```

```
public class RegisterBean implements Serializable{
    private static final long serialVersionUID=1L;
    private String user;
    private String pass;
    private String realname;
    private String gender;
    private Vector chanel;
    public RegisterBean(){
    }
    public String getUser(){
        return user;
    }
    public void setUser(String user){
        this.user=user;
    }
    public String getPass(){
        return pass;
    }
    public void setPass(String pass){
        this.pass=pass;
    }
    public String getRealname(){
        return realname;
    }
    public void setRealname(String realname){
        this.realname=realname;
    }
    public String getGender(){
        return gender;
    }
    public void setGender(String gender){
        this.gender=gender;
    }
    public Vector getChanel(){
        return chanel;
    }
    public void setChanel(Vector chanel){
        this.chanel=chanel;
    }
    public static long getSerialversionuid(){
        return serialVersionUID;
    }
```

```
}
```

3.在项目中新建过滤器类,命名为 Filter。

源代码如下:

```
package myfilter;
import java.io.IOException;
import javax.servlet.FilterChain;
import javax.servlet.FilterConfig;
import javax.servlet.RequestDispatcher;
import javax.servlet.ServletException;
import javax.servlet.ServletRequest;
import javax.servlet.ServletResponse;
public class Filter implements javax.servlet.Filter{
    public void destroy(){
        // TODO Auto-generated method stub
    }
    public void doFilter(ServletRequest request,ServletResponse response,
            FilterChain chain) throws IOException,ServletException{
        try{
            String param_value=this.filterConfig.getInitParameter("filter_off");
            String str=request.getParameter("times");
            if(str==null){
            RequestDispatcher view=request.getRequestDispatcher("login.jsp? param="
            +param_value);
            view.forward(request,response);
            }else
                chain.doFilter(request,response);
        }catch(ServletException e){
            this.filterConfig.getServletContext().log(e.getMessage());
        }catch(IOException e){
            this.filterConfig.getServletContext().log(e.getMessage());
        }catch(Exception e){
            e.printStackTrace();
        }

    }
    private FilterConfig filterConfig;
    public void init(FilterConfig arg0) throws ServletException{
        this.filterConfig=arg0;
    }
}
```

4.新建登录页面 login.jsp。

源代码如下:

```
<%@page language="java" import="java.util.* " pageEncoding="utf-8"%>
<%@page contentType="text/html;charset=utf-8"%>
<html>
    <head>
        <title>login</title>
    </head>
    <body bgcolor="#ffffff">
        <h1>
            登录信息
        </h1>
        <%
            String param=request.getParameter("param");
            String conf=config.getInitParameter("jsp_off");
            if((param!=null)&&(param.equals(conf))){
        %>
        <form method="post" action="register.jsp">
            用户名
            <input type="text" name="user">
            密码
            <input type="password" name="pass">
            <input type="hidden" name="times" value="1">
            <hr>
            <input type="submit" value="提交">
        </form>
        <%
            } else {
        %>
        <jsp:forward page="error.jsp" />
        <%
            }
        %>
    </body>
</html>
```

5.新建注册页面 register.jsp。

源代码如下：

```
<%@page language="java" import="java.util.* " pageEncoding="utf-8"% >
<%@page contentType="text/html;charset=utf-8"% >
<html>
  <head>
    <title>register</title>
  </head>
```

```
<body bgcolor="#ffffff">
<%request.setCharacterEncoding("utf-8");%>
   <h1>学生信息注册</h1>
   <hr>
   <form method="post" action="infodisplay.jsp">
          用户名
          <input type="text" name="user"value="<%=request.getParameter("user")%>"
          readonly="readonly">
          <br>
          密码     
          <input type="password" name="pass" value="<%=
          request.getParameter("pass")%>" readonly="readonly">
          <input type="hidden" name="times" value="1">
          <hr>
          真实姓名<input type="text" name="realname"><br>
          性别      <input type="radio" name=
          "gender" value="男">男   <input type="radio" name=
          "gender" value="女">女
          <br>
          你从哪里知道本网站的主题
          <input type="checkbox" name="chanel" value="网站"/>网站
          <input type="checkbox" name="chanel" value="报纸"/>报纸
          <input type="checkbox" name="chanel" value="电视"/>电视
          <hr><input type="submit" value="提交">
      </form>
  </body>
</html>
```

6.新建显示学生信息页面 infodisplay.jsp。

源代码如下：

```
<%@page language="java" import="java.util.* " pageEncoding="utf-8"
errorPage="error.jsp"%>
<%@page import="mybean.RegisterBean" %>
<%@page contentType="text/html;charset=utf-8"%>
<html>
   <head>
      <title>infodisplay</title>
   </head>
   <body bgcolor="# ffffff">
      <%
         request.setCharacterEncoding("utf-8");
      %>
```

```
<h1>
    学生信息显示
</h1>
<jsp:useBean id="stu" class="mybean.RegisterBean" scope="session">
  <jsp:setProperty property="user" param="user" name="stu" />
  <jsp:setProperty property="pass" param="pass" name="stu"/>
  <jsp:setProperty property="realname" param="realname" name="stu"/>
  <jsp:setProperty property="gender" param="gender" name="stu"/>
</jsp:useBean>
<%
    try{
        Vector vc=new Vector();
        String[] chanel=request.getParameterValues("chanel");
        for(int i=0;i <chanel.length;i++){
            vc.add(chanel[i]);}
        }
        RegisterBean bea=(RegisterBean)session.getAttribute("stu");
        bea.setChanel(vc);
    }catch(Exception e){
        e.printStackTrace();
    }
%>
<hr>
用户名:<jsp:getProperty property="user" name="stu"/><br>
密码:<jsp:getProperty property="pass" name="stu"/><br>
姓名:<jsp:getProperty property="realname" name="stu"/><br>
性别:<jsp:getProperty property="gender" name="stu"/><br>
你从哪里知道本网站的主题:
<%
    try{
        RegisterBean rb=(RegisterBean)session.getAttribute("stu");
        Vector vv=rb.getChanel();
        if(vv!=null){
            for(int i=0;i<vv.size();i++){
                out.println(vv.get(i).toString());
                out.println("   ");
            }
        }
    }catch(Exception e){
        e.printStackTrace();
    }
%>
</body>
```

```
</html>
```

7.新建显示出错提示页面 error.jsp。

源代码如下：

```
<%@page language="java" import="java.util.* " pageEncoding="utf-8"%>
<%@page contentType="text/html;charset=utf-8"%>
<html>
    <head>
        <title>error</title>
    </head>
    <body bgcolor="#ffffff">
        <h1>
            不能直接访问本页面
        </h1>
    </body>
</html>
```

8.配置项目的 web.xml 文件。

源代码如下：

```
<?xml version="1.0" encoding="UTF-8"?>
<web-app version="2.5" xmlns="http://java.sun.com/xml/ns/javaee"
xmlns:xsi="http://www.w3.org/2001/XMLSchema-instance" xsi:schemaLocation=
"http://java.sun.com/xml/ns/javaee  http://java.sun.com/xml/ns/javaee/web-
app_2_5.xsd">
    <filter>
  <filter-name>Filter</filter-name>
  <filter-class>myfilter.Filter</filter-class>
  <init-param>
   <param-name>filter_off</param-name>
   <param-value>200</param-value>
  </init-param>
</filter>
<filter-mapping>
  <filter-name>Filter</filter-name>
  <servlet-name>registerservlet</servlet-name>
</filter-mapping>
<servlet>
  <servlet-name>registerservlet</servlet-name>
  <jsp-file>register.jsp</jsp-file>
</servlet>
<servlet>
  <servlet-name>loginservlet</servlet-name>
  <jsp-file>login.jsp</jsp-file>
  <init-param>
   <param-name>jsp_off</param-name>
```

```
      <param-value>200</param-value>
    </init-param>
  </servlet>
  <servlet-mapping>
    <servlet-name>registerservlet</servlet-name>
    <url-pattern>/register.jsp</url-pattern>
  </servlet-mapping>
  <servlet-mapping>
    <servlet-name>loginservlet</servlet-name>
    <url-pattern>/login.jsp</url-pattern>
  </servlet-mapping>
  <welcome-file-list>
    <welcome-file>register.jsp</welcome-file>
  </welcome-file-list>
  <login-config>
    <auth-method>BASIC</auth-method>
  </login-config>
</web-app>
```

项目部署到 WebLogic 后，运行效果如下：

当未经登录直接访问 infodisplay.jsp 时，会跳转到 error 页面，如图 13-5 所示。

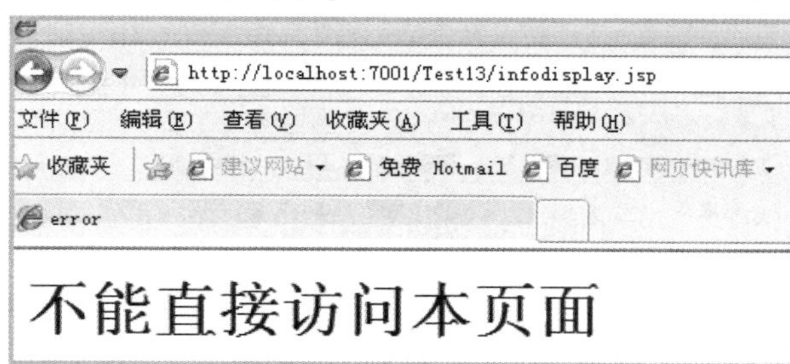

图 13-5 拒绝非法访问

当直接访问注册页面时，如果没有登录，由于过滤器的作用，请求被分配到登录页面，如图 13-6 所示。

图 13-6 过滤器拦截非法访问

当正确输入登录信息后，进入注册页面，如图 13-7 所示。

图 13-7 注册学生信息

填写完注册信息后，单击【提交】按钮，在 infodisplay.jsp 页面中将显示注册信息，效果如图 13-8 所示。

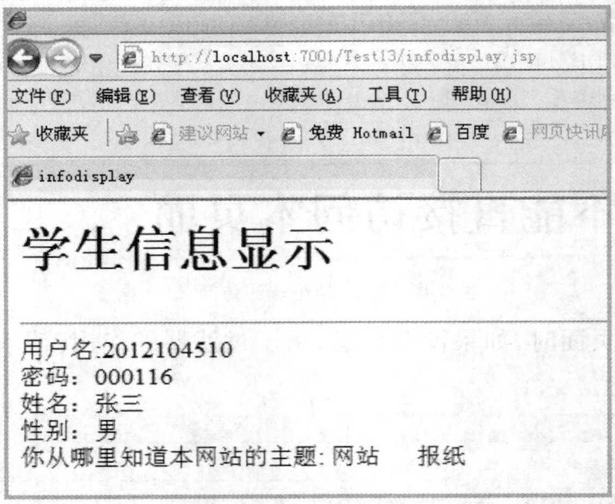

图 13-8 注册后的学生信息

13.5 实验总结

JSP 标准动作利用标记语法格式的标签来控制 Servlet 引擎的行为，利用 JSP 标准动作可以实现动态地插入文件、重用 JavaBean 组件、把用户重定向到另外的页面、为 Java

插件生成 HTML 代码等功能。JSP 标准动作进一步简化了 JSP 程序的开发流程,减少了 JSP 页面上 Java 代码编写的工作量,与 HTML 保持一致,提高了程序的运行效率,是一种在 Web 开发中常用的技术。

13.6 课后思考题

1.什么是 JavaBean,JavaBean 的规范有什么要求?

2.怎样设置或获取 JavaBean 的属性,属性与实例变量是否为同一概念?

3.在 JSP 中使用 JavaBean 有什么要求,使用 JavaBean 有哪两种方法?

4.说明下列标签的功能及其属性的作用:

```
<jsp:useBean id="…" scope="…" class="…" type="…"/>
<jsp:setProperty name="…" property="…" value="…"|param="…"/>
<jsp:getProperty name="…" property="…"/>
```

实验 14

JSP 表达式语言

14.1 实验目的

1.理解什么是 EL
2.熟悉 EL 表达式的基本语法结构
3.熟悉 EL 表达式的特点和使用范围
4.掌握在 JSP 中使用表达式语言的技巧

14.2 实验环境

1.MyEclipse 插件平台
2.WebLogic(或 Tomcat)容器

14.3 实验知识背景

14.3.1 EL 概述

在 JSP 中嵌入大量 Java 代码实现业务逻辑，维护起来较为困难，因此，从 JSP 2.0 起引入了表达式语言 EL。

EL 的最大特点是语法简单、使用方便，可以在模板中直接使用或给标签的属性赋值，能有效减少 JSP 脚本的数量。

EL 的基本格式：${表达式}，如果在 JSP 中需要显示“${”，则需使用“\${”转义。

对应如图 14-1 所示的 JSP 页面的表单信息，分别使用 JSP 小脚本和 EL 表达式来获取这些表单信息。

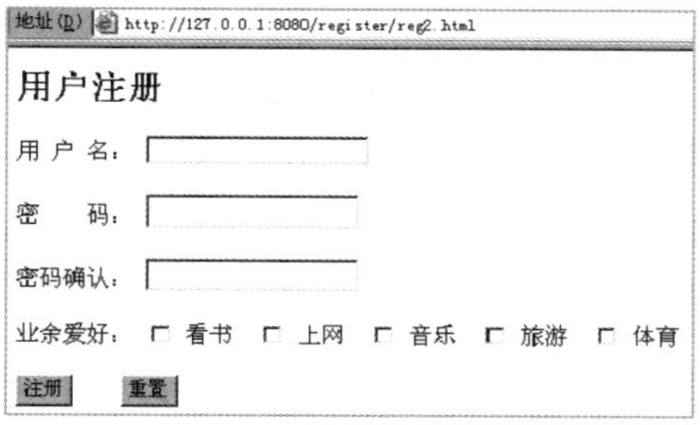

图 14-1 表单样例

使用 JSP,代码如下:

```
//disp1.jsp  使用 JSP 脚本、表达式等
<%request.setCharacterEncoding("GBK");%>
用 户 名:<%=request.getParameter("name")%><p>
密    码:<%=request.getParameter("password")%><p>
密码确认:<%=request.getParameter("rpassword")%><p>
<%
    String[] hobby=request.getParameterValues("hobby");
    String hobbies="";
    for(int i=0;i<hobby.length;i++)
        hobbies=hobbies+ " "+ hobby[i];
%>
业余爱好:<%=hobbies%>
```

如果使用 EL 表达式,代码如下:

```
//disp2.jsp  使用 JSP 的表达式语言(EL)等
<% request.setCharacterEncoding("GBK");% >
用 户 名:${param.name} <p>
密    码:${param.password} <p>
密码确认:${param.rpassword} <p>
业余爱好:${paramValues.hobby[0]} ${paramValues.hobby[1]}
${paramValues.hobby[2]} ${paramValues.hobby[3]}${paramValuse.hobby[4]}
```

由上面的例子可知,使用 EL 表达式和使用 JSP 表达式、脚本相比,代码更简洁,使用上更直观。

EL 的功能主要有:

(1)精确访问作用域变量,即属性 attribute

(2)简单的 JavaBean 访问语法

(3)集合元素的简洁表示

(4)对 request 参数和 Cookie 的简单访问

(5)简单的运算符

(6)条件表达式求值

(7)自动类型转换

(8)使用空值代替 NullPointerException

要在 JSP 页面中使用 EL,需要使用 page 指令进行指定,格式为:

<%@page isELIgnored=true|false%>

其中,true 值表示不能解析 EL,false 值则表示可以解析 EL。从 JSP 2.0 起默认支持 EL,通常可不设置。

14.3.2 EL 基础语法

1.EL 中的常量

(1)布尔常量:只有 true 和 false 两个值。

(2)整数常量:同 Java 中的整数,范围为 Long.MIN_VALUE~Long.MAX_VALUE。

(3)浮点常量:同 Java 中的浮点数,范围为 Double.MIN_VALUE~Double.MAX_VALUE。

(4)字符串常量:用双引号("")和单引号('')括起来的一串字符。只有与分界符相同时,才需要进行转义(\"或\')。

(5)null 常量:只有一个,即 null。

2.EL 中的变量

EL 将变量映射到一个对象上,其中变量不用预先定义,如果是隐式对象,则直接使用;若为非隐式对象,则依次在 page、request、session、application 中查找,若找不到,则返回 null。

3.EL 对出错信息的处理

EL 对错误信息的处理更为友好。例如:user 对象不存在,${user}返回为空(非 null),即使是 ${user.name}也为空,同样不会抛出异常。但如果变量存在,而属性不存在,则抛出异常。

4.自动转换类型

EL 提供的另一个功能是自动转变类型,例如:${param.count+20},假若表单传来的 count 值为 10,那么上面的结果为 30。如果不使用 EL 表达式,要得到这一结果,需要如下代码:

```
String str_count=request.getParameter("count");
int count=Integer.parseInt(str_count);
count=count+20;
```

14.3.3 EL 运算符

1.点号"."和方括号"[]"

点号".":通常用于访问对象的属性,例如:${user.userName}访问 user 对象的 userName 属性;再如:${user.userName.firstName}。如果通过 JSP 脚本实现同一功能,可以使用这种形式的语句:user.getUserName().getFirstName()。

方括号“[]”：通常用于访问数组或集合的元素，对于实现了 java.util.Map 接口的集合，方括号运算符使用关联的键查找存储在映射中的值，例如：${paramValues.hobby[0]}。

通常，点号“.”和方括号“[]”可以互用，例如：${sessionScope.user.sex} 等同于 ${sessionScope.user["sex"]}；也可以混用，如：${sessionScope.shoppingCart[0].price}。不过，在下列两种情况下，两者之间存在差异：

(1)当要存取的属性名中包含一些特殊字符，如“.”或“-”等并非字母或数字的符号，就一定要使用“[]”，例如：

${user.My-Name}写法不正确，应改为：${user["My-Name"]}。

(2)方括号内的值是一个变量时，例如：

${sessionScope.user[data]}，此时 data 是一个变量，假若 data 的值为“sex”时，那么上一语句等同于 ${sessionScope.user.sex}；假若 data 的值为“name”时，则等同于 ${sessionScope.user.name}。但是，点号“.”无法动态取值。

2.EL 中的算术运算符

EL 中算术运算符有 5 个，包括：+、-、*、/(或 div)和 %(或 mod)，除法中即使是“5/0”也不会抛出异常，而是友好显示“infinity”。

3.EL 中的比较运算符

EL 中的比较运算符有 6 个，包括：==(或 eq)、!=(或 ne)、<(或 lt)、>(或 gt)、<=(或 le)和 >=(或 ge)。

4.EL 中的逻辑运算符

EL 中的逻辑运算符有 3 个，包括：&&(或 and)、||(或 or)和 !(或 not)。

注意：在使用 EL 关系运算符时，不能写成：

```
${param.password1}== ${param.password2}
```

或者

```
${ ${param.password1}== ${param.password2}}
```

而应写成：

```
${param.password1==param.password2 }
```

5.验证运算符

EL 中的验证运算符为 empty。例如：${empty user}。

6.条件运算符

格式：条件?值 1:值 2

例如：${sex=="1"?"男":"女"}。

7.圆括号( )

该符号可以改变运算顺序。

14.3.4　EL 的隐式对象

1.pageContext 对象

与 JSP 中的 pageContext 对象相同，可以获取其他对象，如表 14-1 所示。

表 14-1　　pageContext 对象的使用说明表

表达式格式	说明
${pageContext.request.queryString}	取得请求中的参数字符串
${pageContext.request. requestURL}	取得请求的 URL，但不包括请求的参数字符串
${pageContext.request.contextPath}	取得服务的 Web application 的名字

2.与范围有关的隐式对象

(1)pageScope：与“页面”作用域属性的名称和值相关联的 Map 类。

(2)requestScope：与“请求”作用域属性的名称和值相关联的 Map 类。

(3)sessionScope：与“会话”作用域属性的名称和值相关联的 Map 类。

(4)applicationScope：与“应用程序”作用域属性名称和值相关联的 Map 类。

以上隐式对象与 JSP 的 pageContext、request、session 和 application 一样，但是，这 4 个隐式对象只能用来取得范围属性值，即相当于 JSP 中的 getAttribute(String name)，却不能取得其他相关信息。而 JSP 中的隐式对象除获取属性外，还可以有其他用途，例如：JSP 中的 request 对象除可以存取属性之外，还可以取得用户的请求参数或表头信息等。但是在 EL 中，它就只能用来取得对应范围的属性值，例如：

${sessionScope.username}可获取 session 中名为 username 的属性值，相当于 JSP 中的 session.getAttribute("username")。

在实际操作时，可省略范围隐式对象，如：${. username}。这样，会依次在 pageScope、requestScope、sessionScope、applicationScope 中查找。

3.与输入有关的隐式对象

(1)param：按名称存储请求参数单一值的 Map 类。

(2)paramValues：请求参数的所有值构成的 String 数组。

例如：

${param.name} 相当于：request.getParameter("name")。

${paramValues.hobby} 相当于：request.getParameterValues("name")。

4.其他隐式对象

(1)与请求头有关的隐式对象共 2 个。

①header：请求头参数值的 Map 类，例如：${header.host}。

②header Values：请求头参数的所有值构成的 String 数组。

(2)获取 Web 应用程序初始参数的隐式对象 initParamter。

例如：${initParam.userid}相当于：

```
String userid=(String)application.getInitParameter("userid");
```

(3)用来获得名为 userid 的 Web 上下文参数。

获取 Cookie 对象的隐式对象：Cookie

例如：要取得 Cookie 中 color 的值，可以使用 ${cookie.color}来获得。

14.4 实验内容与步骤

按要求编写 JSP 程序，使用 EL 表达式来简化订单生产页面的设计，功能需求如下：

(1)使用 checkOut.jsp 接收用户的信息，主要信息有客户姓名、送货地址、邮政编码、联系电话和移动电话，将这些信息提交后发送到 middle.jsp。

(2)middle.jsp 将从 checkOut.jsp 提交过来的表单数据封装在 OrderBean 对象中。

(3)从 OrderBean 中读取数据，并用 EL 表达式显示在页面中。

1.新建 Web 项目，项目结构如图 14-2 所示。

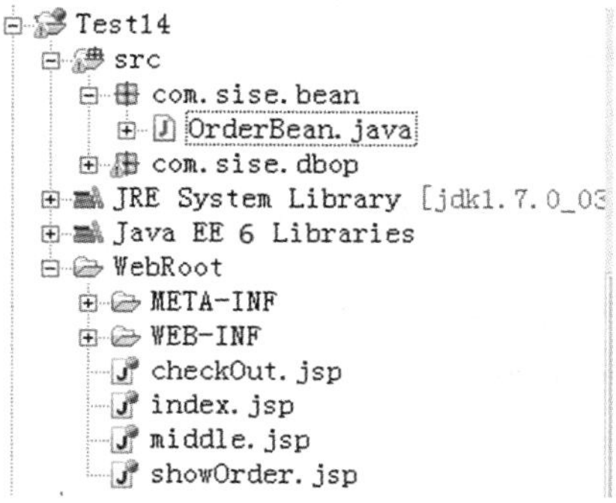

```
Test14
  src
    com.sise.bean
      OrderBean.java
    com.sise.dbop
  JRE System Library [jdk1.7.0_03
  Java EE 6 Libraries
  WebRoot
    META-INF
    WEB-INF
    checkOut.jsp
    index.jsp
    middle.jsp
    showOrder.jsp
```

图 14-2　订单生成项目结构图

2.在项目中新建 JavaBean 类，命名为 OrderBean。

源代码如下：

```
package com.sise.bean;
public class OrderBean{
    private String customerName;// 客户姓名
    private String address;// 地址
    private String zipCode;// 邮政编码
    private String telephone;// 联系电话
    private String movePhone;// 移动电话
    private String notice;// 订单附言
    private double totalPrice;// 付款金额
    /**
     *@return totalPrice 返回付款金额
     * /
    public double getTotalPrice(){
        return totalPrice;
    }
    /**
     * param totalPrice 设置付款金额
     * /
    public void setTotalPrice(double totalPrice){
        this.totalPrice=totalPrice;
```

```
}
/**
 *@return address 返回地址
 */
public String getAddress(){
    return address;
}
/**
 * param address 设置地址
 */
public void setAddress(String address){
    this.address=address;
}
/**
 *@return customerName 返回客户姓名
 */
public String getCustomerName(){
    return customerName;
}
/**
 * param customerName 设置客户姓名
 */
public void setCustomerName(String customerName){
    this.customerName=customerName;
}
/**
 *@return movePhone 返回移动电话
 */
public String getMovePhone(){
    return movePhone;
}
/**
 * param movePhone 设置移动电话
 */
public void setMovePhone(String movePhone){
    this.movePhone=movePhone;
}
/**
 *@return notice 设置订单附言
 */
public String getNotice(){
    return notice;
}
```

```
    /**
     * param notice 返回订单附言
     */
    public void setNotice(String notice){
        this.notice=notice;
    }
    /**
     *@return telephone 返回联系电话
     */
    public String getTelephone(){
        return telephone;
    }
    /**
     * param telephone 设置联系电话
     */
    public void setTelephone(String telephone){
        this.telephone=telephone;
    }
    /**
     *@return zipCode 返回邮政编码
     */
    public String getZipCode(){
        return zipCode;
    }
    /**
     * param zipCode 设置邮政编码
     */
    public void setZipCode(String zipCode){
        this.zipCode=zipCode;
    }
}
```

3.新建页面 checkOut.jsp。

源代码如下：

```
<%@page language="java" pageEncoding="gb2312"%>
<%@page import="java.text.* "%>
<HTML>
<HEAD>
<TITLE>提交订单</TITLE>
</HEAD>
<BODY leftMargin=0 topMargin=0 marginheight="0" marginwidth="0">
<TABLE cellSpacing=0 cellPadding=0 width=776 align=center border=0>
    <TBODY>
        <TR valign=top>
```

```
            <TD>
            </TD>
            <TD>
<TABLE cellSpacing=0 cellPadding=0 width="96% " align=center border=0>
                <TBODY>
                    <TR>
                        <TD><BR>
                        </TD>
                    </TR>
                </TBODY>
            </TABLE>
            <!-- 向 Servlet - OrderServlet 提交 POST 请求 -->
            <form method="POST" action="/Test14/middle.jsp">
            <table cellspacing=1 cellpadding=4 width="92% " border=0
                align="CENTER" bgcolor="#c0c0c0">
                <tr bgcolor="#dadada">
                    <td colspan="5" height="25" align=center>
<font color="#000000">请确认支付和配送信息</font></td>
                </tr>
                <!-- 提交"客户姓名"信息 -->
                <tr bgcolor="#ffffff">
                    <td width="22%" align="RIGHT"><font color="#000000">
                        客户姓名:</font></td>
                    <td colspan=4 width="78%"><input type="text" name="customerName"
                      size="46" maxlength="20"></td>
                </tr>
                <!-- 提交"送货地址"信息 -->
                <tr bgcolor="#ffffff">
                    <td width="22%" align="RIGHT"><font color="#000000">
                        送货地址:</font></td>
                    <td colspan=4 width="78%"><input type="text" name="address"
                        size="46" maxlength="200"></td>
                </tr>
                <!-- 提交"邮政编码"信息 -->
                <tr bgcolor="#ffffff">
                    <td width="22%" align="RIGHT"><font color="#000000">
                        邮政编码:</font></td>
                    <td colspan=4 width="78%"><input type="text" name="zipCode"
                    size="46" maxlength="6"></td>
                </tr>
                <!--提交"联系电话"信息 -->
                <tr bgcolor="#ffffff">
                    <td width="22% " height="31" align="RIGHT"><font
```

```
                color="#000000">联系电话:</font></td>
                <td colspan=4 width="78% " height="31"><input type="text"
                name="telephone" size="46" maxlength="13"></td>
            </tr>
            <!-- 提交“移动电话”信息 -->
            <tr bgcolor="#ffffff">
                <td width="22%" height="31" align="RIGHT"><font
                color="#000000">移动电话:</font></td>
                <td colspan=4 width="78%" height="31"><input type="text"
                name="movePhone" size="46" maxlength="12"></td>
            </tr>
                <tr bgcolor="#dadada">
                 <!-- 提交信息 -->
                <td colspan="5" height="12" align="center">
                <input type=submitvalue="确认以上信息无误,提交" name=Submit
                ></td>
            </tr></table>
        </form></TD></TR></TBODY>
</TABLE>
</BODY>
</HTML>
```

4.新建页面 middle.jsp。

源代码如下:

```
<%@page language="java" pageEncoding="gb2312"%>
<%response.setContentType("text/html;charset=gb2312");%>
<%request.setCharacterEncoding("gb2312");%>
<jsp:useBean id="orderBean" class="com.sise.bean.OrderBean" scope="session"/>
<!-- 将从 checkOut.jsp 提交过来的表单数据用 EL 封装在 OrderBean 对象中 -->
<jsp:setProperty name="orderBean" property="customerName" value=
"${param.customerName}"/>
<jsp:setProperty name="orderBean" property="address" value="${param.address}"/>
<jsp:setProperty name="orderBean" property="zipCode" value="${param.zipCode}"/>
<jsp:setProperty name="orderBean" property="telephone" value="${param.telephone}"/>
<jsp:setProperty name="orderBean" property="movePhone" value="${param.movePhone}"/>
<!-- 自动跳转到 showOrder.jsp -->
<jsp:forward page="showOrder.jsp"/>
<html>
    <head>
        <title>订单信息显示跳转</title>
    </head>
    <body></body>
</html>
```

5.新建订单显示页面 showOrder.jsp。

源代码如下：

```
<%@page language="java" pageEncoding="gb2312"%>
<!-- 设置输入编码为中文 -->
<%response.setContentType("text/html;charset=gb2312");%>
<%request.setCharacterEncoding("gb2312");%>
<!-- 使用 OrderBean.java -->
<jsp:useBean id="orderBean" class="com.sise.bean.OrderBean" scope="session"/>
<html>
<head>
<title>订单信息显示</title>
</head>
<body>
<table width="316" border="1" >
  <tr>
    <th colspan="2" bgcolor="# 999999" scope="col"><span class="STYLE2">您提交
    的订单信息如下:</span></th>
  </tr>
  <!-- 从 OrderBean 中读取数据,并用 EL 显示在页面中 -->
  <tr>
    <td width="90">客户姓名:</td>
    <td width="240"> ${orderBean.customerName}</td>
  </tr>
  <tr>
    <td>送货地址:</td>
    <td> ${orderBean.address}</td>
  </tr>
  <tr>
    <td>邮政编码:</td>
    <td> ${orderBean.zipCode}</td>
  </tr>
  <tr>
    <td>联系电话:</td>
    <td> ${orderBean.telephone}</td>
  </tr>
  <tr>
    <td>移动电话:</td>
    <td> ${orderBean.movePhone}</td>
  </tr>
</table>
</body>
</html>
```

项目部署到 WebLogic 并运行。在 checkOut.jsp 页面提交订单信息，如图 14-3 所示。

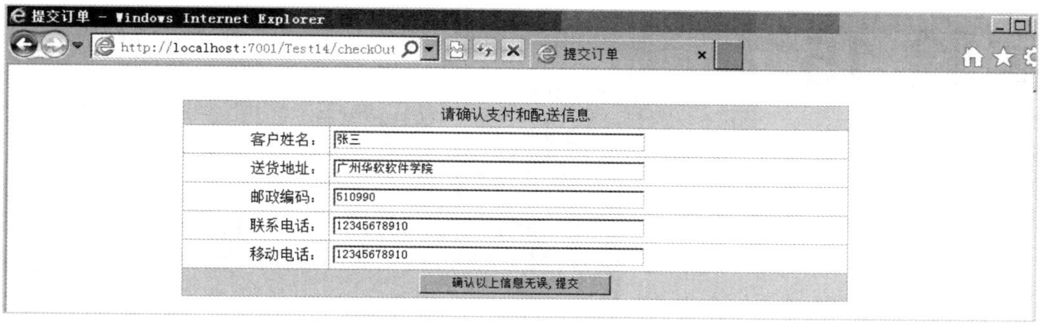

图 14-3 提交订单信息

使用 EL 表达式在 showOrder.jsp 页面中将订单内容显示，如图 14-4 所示。

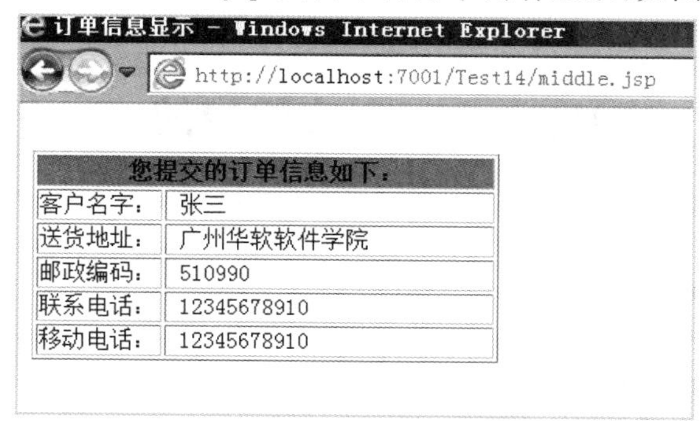

图 14-4 显示订单信息

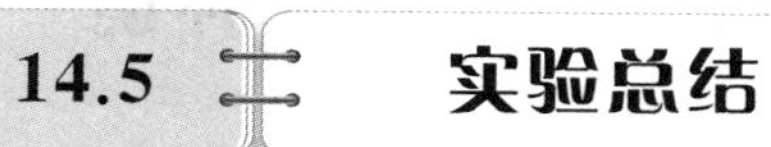

14.5 实验总结

本次实验主要练习了 JSP 2.0 中的 EL 语法及其应用。通过本章的学习，在后续的开发中，JSP 页面中可以减少很多不必要的 Java 代码，让程序的可读性得以提高，从而解决了使用 JSP 标准动作只能处理单一业务的缺陷，使得业务逻辑和表现逻辑分离工作更加轻松。

14.6 课后思考题

1.简述 EL 的格式、特点和使用环境。

2.说明下列表达式的功能或结果：

(1) ${sessionScope.user.name}

(2) ${sessionScope.user["password"]}

(3) ${(content==null)?"内容为空":content}

(4) ${.username}

(5) ${10/0}

3.若要获得用户的 IP 地址，怎样用表达式来书写？

实验 15

JSP 标准标签库

15.1 实验目的

1.理解什么是 JSTL
2.掌握 JSTL 的配置方法
3.掌握核心标签库的使用技巧
4.掌握国际化和格式化标签库的使用技巧

15.2 实验环境

1.MyEclipse 插件平台
2.WebLogic(或 Tomcat)容器

15.3 实验知识背景

15.3.1 JSTL 简介

JSTL (JSP Standard Tag Library,JSP 标准标签库)是一个开源项目,是一个标准的已定制好的 JSP 标签库。JSTL 可以替代 Java 代码实现各种功能,如输入输出、流程控制、迭代、数据库查询及国际化的应用等,可减少 JSP 中脚本代码的数量。

下载 jakarta-taglibs-standard-1.2.1.zip,解压后得到两个文件:jstl.jar 和 standard.jar,其中,jstl.jar 包含的是 JSTL 规范中定义的接口和类,standard.jar 包含的是 Jakarta 小组对 JSTL 的实现和 JSTL 中 5 个标签库的 TLD 文件。将上述两个文件直接复制到应用程序的 WEB-INF\lib 目录下即可。在 MyEclipse 等开发工具中,已包含 JSTL 内容,不必下载和安装,只需在创建 Web 项目时选择 JSTL 即可,如图 15-1 和图 15-2 所示。

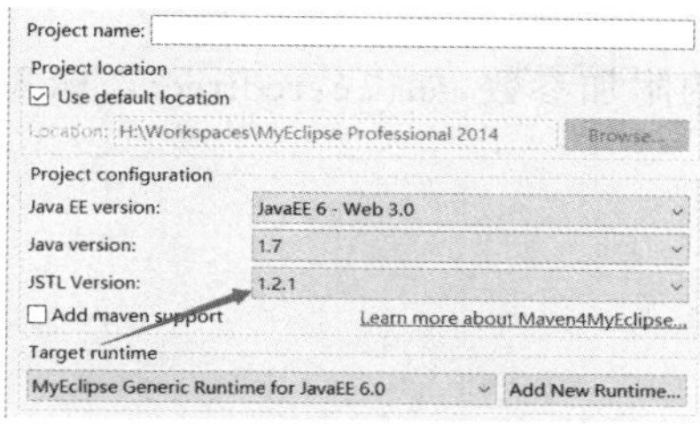

图 15-1　选择 JSTL 版本

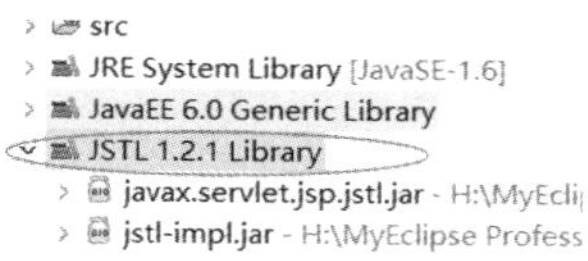

图 15-2　添加包的引用

15.3.2　JSTL 语法基础

核心标签库通用格式:

```
<%@taglib prefix="c" uri="http://java.sun.com/jsp/jstl/core"%>
<c:xxx 属性 1="值 1" … 属性 k="值 k"…>
```

下面对常用的核心标签库的标签进行介绍。

1.<c:out>:主要用来显示数据的内容,与<%=表达式%>等效。

格式 1:没有标签体

```
<c:out value="value"[escapeXml="true|false"][default="默认值"]/>
```

格式 2:有标签体

```
<c:out value="value"[escapeXml="true|false"]>
  默认值
</c:out>
```

2.<c:set>:用来将变量存储在 JSP 范围中或 JavaBean 的属性中。

格式 1:将 value 的值储存在 scope 范围内的 varName 变量中

```
<c:set value="value" var="varName"[scope="page|request|session|application"]/>
```

或

```
<c:set var="varName" [scope="page|request|session|application"]>
  标签体内容
</c:set>
```

格式 2:将 value 的值保存至 target 对象的属性中

```
<c:set value="value" target="对象名" property="propertyName" />
```

或

```
<c:set target="对象名" property="propertyName">
  标签体内容
</c:set>
```

3.<c:if>条件判断

格式:

```
<c:if test="测试条件" var="varName"[scope="page|request|session|application"]/>
```

或

```
<c:if test="测试条件" [var="varName"] [scope="page|request|session|application"]>
```

```
  具体内容
</c:if>
```

4.<c:param>:为其他标签提供 URL 的附加参数,如<c:redirect>标签。

格式:

```
<c:param name="名字" value="值" />
```

或

```
<c:param name="名字" >
     参数值
</c:param>
```

5.<c:redirect>:实现 URL 跳转。

格式:

```
<c:redirect url="url" [context="context"] />
```

或

```
<c:redirect url="url" [context="context"]>
   <c:param>(代表查询字符串(Query String)参数)
</c:redirect>
```

6.多分支判断:<c:choose>、<c:when>和<c:otherwise>。

格式:

```
<c:choose>
    <c:when test="条件表达式">
       body
    </c:when>
    <c:when test="条件表达式">
       body
    </c:when>
       …
       [<c:otherwise>
       body
     </c:otherwise>]
</c:choose>
```

7.<c:forEach>:用于对一个集合中的元素进行循环迭代操作,或按指定的次数重复迭代执行标签体中的内容。

格式 1:在集合对象中迭代

```
<c:forEach [var="变量名"]items="集合名" [varStatus="迭代信息状态"][begin=
"起始索引"][end="终止索引"][step="步长"]>
   标签体内容
</c:forEach>
```

格式 2:按指定次数进行迭代

```
<c:forEach[var="变量名"][varStatus="迭代信息状态"]begin="起始索引" end="终止索引" [step
="步长"]>
```

```
    标签体内容
</c:forEach>
```

8.<c:forTokens>:用来浏览一个字符串中所有的成员,由定义符号(delimiters)分隔。

格式:

```
<c:forTokens items="迭代对象" delims="分隔符" [var="varName"]
    [varStatus="varStatusName"] [begin="begin"] [end="end"] [step="step"]>
      标签体内容
</c:forTokens>
```

15.4 实验内容与步骤

1.问题描述:用 JSTL 与 EL 技术实现用户注册后显示注册信息的功能。图 15-3 为用户注册页面,用户单击【注册】按钮后,则提取用户输入信息,并将输入信息进行显示,运行效果如图 15-4 所示。

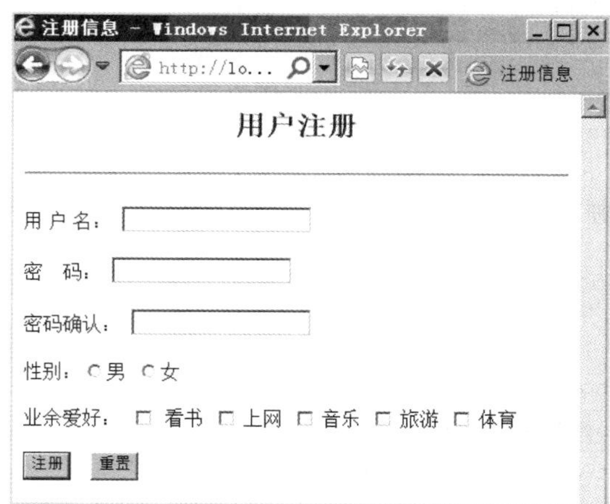

图 15-3　用户注册页面

图 15-4　注册信息显示页面

(1)新建 Web 项目,先把 jstl.jar 和 standard.jar 拷贝到 WEB-INF/lib 目录中,然后

添加引用，项目结构如图 15-5 所示。

```
Test15
  src
  JRE System Library [jdk1.7.0_03
  Java EE 5 Libraries
  Web App Libraries
  Referenced Libraries
    jstl.jar
    standard.jar
  WebRoot
    META-INF
    WEB-INF
      lib
        jstl.jar
        standard.jar
      web.xml
    disp.jsp
    order.jsp
    orderR.jsp
    reg.jsp
```

图 15-5　用户注册项目结构图

(2)在项目中新建页面 reg.jsp。

源代码如下：

```
<%@page language="java" import="java.util.* " pageEncoding="utf-8"%>
<html>
    <head>
        <title>注册信息</title>
    </head>
    <body>
    <div align=center>
        <H2>
            用户注册
        </H2>
    </div>
        <hr/>
        <form method="post" action="disp.jsp">
            用  户  名：
            <input type="text" name="name">
            <P>
                密     码：
                <input type="password" name="password">
            <P>
                密码确认：
                <input type="password" name="rpassword">
            <P>
            性别：<input type="radio" name="sex" value="1"/>男
            <input type="radio" name="sex" value="0"/>女
            <p/>
                业余爱好：
                <input type="checkbox" name="hobby" value="看书">
                看书  
```

```
            <input type="checkbox" name="hobby" value="上网">
            上网  
            <input type="checkbox" name="hobby" value="音乐">
            音乐  
            <input type="checkbox" name="hobby" value="旅游">
            旅游  
            <input type="checkbox" name="hobby" value="体育">
            体育  
        <P>
            <input type="submit" name="Submit" value="注册">
              
            <input type="reset" name="Reset" value="重置">
        <P>
      </form>
   </body>
</html>
```

(3)新建页面 disp.jsp。

源代码如下:

```
<%@page language="java" import="java.util.* " pageEncoding="utf-8"%>
<%@taglib prefix="c" uri="http://java.sun.com/jsp/jstl/core"%>
<%request.setCharacterEncoding("utf-8");%>
<html>
   <head>
      <title>EL+ JSTL 例子</title>
   </head>
   <body>
         用户姓名:${param.name}
      <p>
         用户密码:${param.password}
      <p>
         确认密码:${param.password}
      <p>
         性别:${param.sex=="0"?"女":"男"}
      <p>
         业余爱好:
      <c:forEach items="${paramValues.hobby}" var="hobby">
         ${hobby}
      </c:forEach>
   </body>
</html>
```

2.问题描述:在上一个项目中,继续编写两个 JSP 页面,要求可以实现用表格显示商品名字和单价，允许用户输入购买的数量，如图 15-6 所示。当用户单击【提交】按钮后，则提取用户输入数据，进行运算，求出货品的总价并且输出。

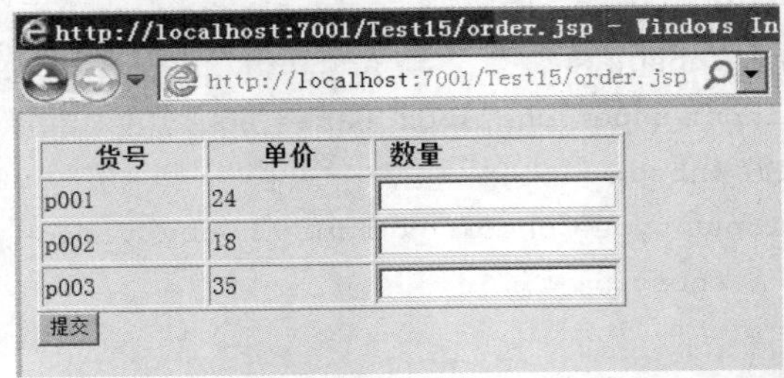

图 15-6　显示商品信息

(1)在项目中新建页面 order.jsp。

源代码如下：

```
<%@page contentType="text/html;charset=gb2312"%>
<%@page import="java.sql.*"%>
<HTML>
   <BODY bgcolor=#eeeeee>
      <Font size=3>
         <form action=orderR.jsp method=post>
            <Table Border>
               <TR>
                  <TH width=100>
                     货号<TH>
                     <TH width=100>
                        单价<TH>
                        <TH width=50>
                           数量<TH>
               </TR>
                  <TR>
                  <TD><input type="hidden" name="No" value="p001"/>
                     p001
                  </TD>
                  <TD><input type="hidden" name="price" value="24"/>
                     24
                  </TD>
                  <TD><input type="text" name="num" />
                  </TD>
               </TR>
               <TR>
                  <TD>
                     <input type="hidden" name="No" value="p002"/>
                     p002
                  </TD>
```

```
                    <TD><input type="hidden" name="price" value="18"/>
                        18</TD>
                    <TD><input type="text" name="num" value=""/>
                </TD>
                </TR>
                <TR>
                    <TD><input type="hidden" name="No" value="p003"/>
                        p003
                    </TD>
                    <TD><input type="hidden" name="price" value="35"/>
                        35
                    </TD>
                    <TD><input type="text" name="num" value=""/>
                </TD>
                </TR>
            </Table>
            <input type="submit" value="提交">
        </form>
    </BODY>
</HTML>
```

(2)在项目中新建页面 orderR.jsp。

源代码如下：

```
<%@page contentType="text/html;charset=gb2312"%>
<%@taglib prefix="c" uri="http://java.sun.com/jsp/jstl/core"%>
<HTML>
    <BODY bgcolor="#eeeeee">
        <Font size=3>
            <table border>
                <tr>
                    <th>货号</th>
                    <th>单价</th>
                    <th>数量</th>
                    <th>总价</th>
                </tr>
            <c:forEach items="${paramValues.No}" varStatus="s">
                <tr>
                    <td> ${paramValues.No[s.index]}</td>
                    <td> ${paramValues.price[s.index]}</td>
                    <td> ${paramValues.num[s.index]}</td>
                    <td>
                    ${paramValues.price[s.index]* paramValues.num[s.index]}
                    </td>
                </tr>
```

```
                </c:forEach>
                </table>
            </FONT>
        </BODY>
    </HTML>
```

将项目部署到 WebLogic 并运行。

先输入货物数量，如图 15-7 所示。

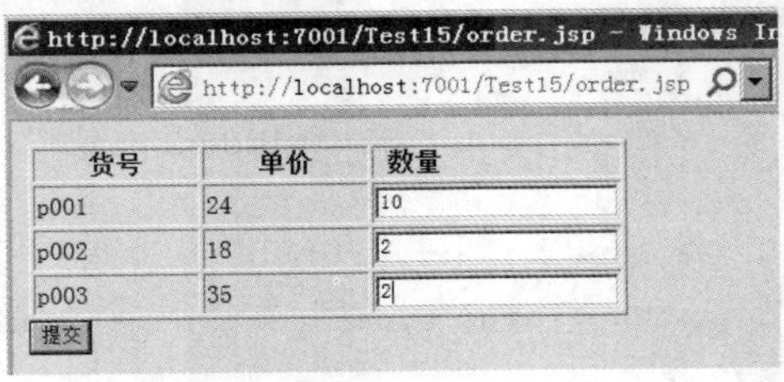

货号	单价	数量
p001	24	10
p002	18	2
p003	35	2

图 15-7　输入购买数量

单击【提交】按钮，计算出货物的总价，如图 15-8 所示。

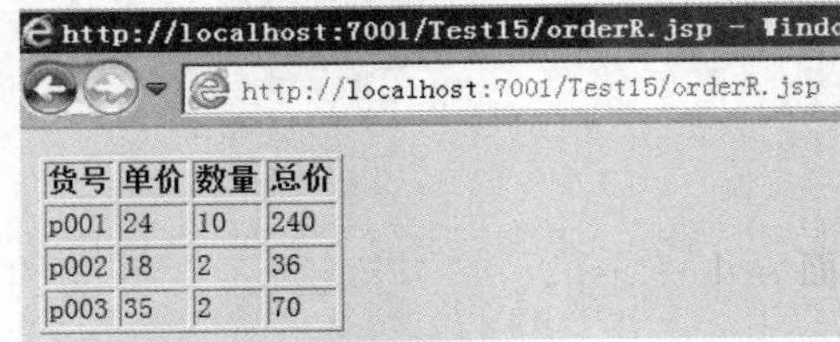

货号	单价	数量	总价
p001	24	10	240
p002	18	2	36
p003	35	2	70

图 15-8　计算货物总价

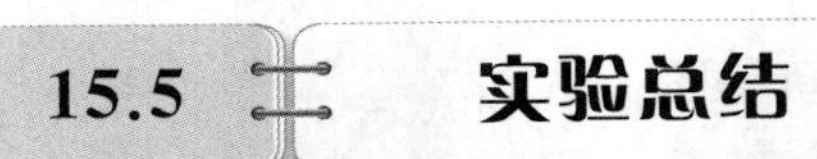

15.5 实验总结

本次实验主要练习常用的 JSTL 标签的使用。JSTL 结合前面学习的 EL 表达式，可以进一步简化 Web 程序的开发流程，使非 Java 程序员也能很快上手，实现快速的动态页面开发。

15.6 课后思考题

1.要使用 JSTL，需要包含哪两个 jar 包，它们安装在什么位置？

2.使用 JSTL 时，相应的 taglib 指令应如何书写？

3.标签<c:out>的功能是什么，属性 value、escapeXml、default 分别有什么含义？

4.标签<c:set>的功能是什么，属性 var、value、scope、target、property 各有什么含义？

实验 16 自定义 JSP 标签

16.1 实验目的

1.理解自定义标签体系结构
2.掌握创建自定义标签的基本步骤
3.掌握给自定义标签添加属性的方法
4.掌握用自定义标签处理标签体的方法

16.2 实验环境

1.MyEclipse 插件平台
2.WebLogic(或 Tomcat)容器

16.3 实验知识背景

16.3.1 自定义标签库简介

在 JSP 中使用标签可以减少 Java 代码的编写,便于页面的维护。但由于 JSP 的标准操作和 JSTL 定义的标签不能满足用户的所有要求,在 JSP 2.0 中允许用户根据需要创建自己的标签库,即自定义标签库。

用户自定义标签和 JSTL 中的标签从技术上看没有任何区别,可以将这些标签统称为 JSP 标签。JSP 标签在 JSP 页面中通过 XML 语法格式被调用,当 JSP 引擎将 JSP 页面翻译成 Servlet 时,就将这些调用转换成相应的 Java 代码。本质上,JSP 标签调用了部分 Java 代码,只是这些 Java 代码在 JSP 页面中以另外一种形式(XML 语法格式)表现出来。

16.3.2 自定义标签的形式

自定义标签的形式有四种：空标签、带有属性的空标签、带有标签体的标签、带有标签体和属性的标签。

1.空标签(不含标签体和属性)

格式：<前缀:标签名/> 或 <前缀:标签名></前缀:标签名>

例如：

```
<simple:greeting/>
```

2.带有属性的空标签

格式一：<前缀:标签名 属性 1="值 1" 属性 2="值 2"…/>

格式二：<前缀:标签名 属性 1="值 1" 属性 2="值 2"…>

</前缀:标签名>

例如：

```
<simple:greetingAtt name="<%=username%>"/>
```

3.带有标签体的标签

格式：<前缀:标签名>

标签体

</前缀:标签名>

例如：

```
<simple:greetingBodyTag>
    <%=hr%>:<%=min%>:<%=sec%>
    </simple:greetingBodyTag>
```

4.带有标签体和属性的标签

格式：<前缀:标签名 属性 1="值 1" 属性 2="值 2"…>

标签体

</前缀:标签名>

例如：

```
<simple:greetingAtt name="<%=username%>" >
    现在时间是:<%=hr%>:<%=min%>:<%=sec%>
  </simple:greetingAtt>
```

要注意的是，一个自定义标签包含起始标记和结束标记；在起始标记中可设置自定义标签属性；在自定义标签的起始标记和结束标记之间还可以有标签体；自定义标签按照以下顺序执行：起始标记→标签体→结束标记。

16.3.3 自定义标签的工作原理及相关概念

自定义标签实际上是一个实现了特定接口的 Java 类，封装了一些常用功能，在运行时被相应的代码所替换。如图 16-1 所示。

为了方便后续的学习，下面给出一些概念的定义。

标签(Tag)：标签是一种 XML 元素，其名称和属性均对大小写敏感。通过标签可以让 JSP 页面实现特定功能，使得 JSP 页面变得简洁并且易于维护。

标签库(Tag Library)：由一系列功能相似、逻辑上互相联系的标签构成的集合，同一

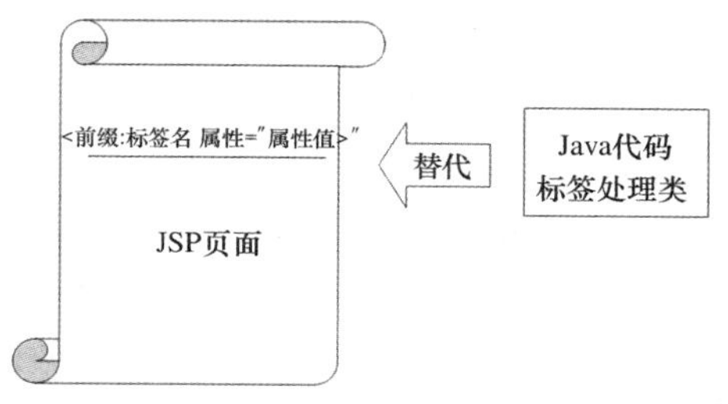

图 16-1　自定义标签的工作原理示意图

个标签库的前缀相同。

标签库描述文件(Tag Library Descriptor):是一个 XML 文件,提供了标签库中类和 JSP 中对标签引用的映射关系,也是一个配置文件,与 web.xml 类似。

标签处理类(Tag Handle Class):是一个 Java 类,这个类继承了 TagSupport(或 BodyTagSupport)类,也可以实现 Tag 或其子接口,通过该类可以自己定义 JSP 标签的具体功能。

16.3.4　自定义标签相关 API

自定义标签相关的类与接口位于 javax.servlet.jsp.tagext 包中,具体结构如图 16-2 所示。

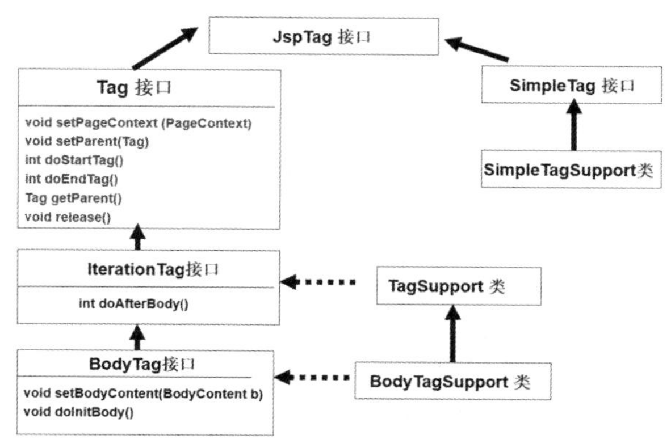

图 16-2　自定义标签相关的类与接口

说明:

1.JSP Tag 是 Tag 和 SimpleTag 接口的父接口,是一个标记接口,不包含任何属性和方法。

2.Tag 接口中的方法和常量:

(1)setPageContext()方法:负责设置页面的 pageContext。

(2)setParent()方法:负责设置父标签。

(3)doStartTag()方法:遇到自定义标签的开始标记时去调用标签处理类的方法,返回值为 EVAL_BODY_INCLUDE(表示标签体要执行,执行结果放在当前输出流中)或 SKIP_BODY(不执行标签体)。

(4)doEndTag()方法：遇到自定义标签的结束标记时去调用标签处理类的方法，返回值为 EVAL_PAGE(JSP 页面的剩余内容将继续执行)或 SKIP_PAGE(JSP 页面的剩余内容不执行)。

(5)doAfterBody()方法：是 IterationTag 接口增加的方法，在执行完标签体后调用，如果没有标签体，该方法将不会调用。该方法的返回值是 SKIP_BODY 或 EVAL_BODY_AGAIN(重复执行标签体)。

(6)setBodyContent()方法：是 BodyTag 接口中设置 bodyContent 属性的方法，以备后面获取标签体内容；只有在 doStartTag()返回 EVAL_BODY_BUFFERED 时才执行。

(7)doInitBody()方法：在 setBodyContent()方法之后、执行标签体之前调用，为标签体的执行做准备；只有在 doStartTag()返回 EVAL_BODY_BUFFERED 时才执行。

3.TagSupport 类实现 IterationTag 接口，使用它可简化标签处理类的创建。

4.BodyTagSupport 类实现 BodyTag 接口，使用它可简化有关标签体处理类的创建。

16.4 实验内容与步骤

1.创建<simple:greeting />标签，该标签的功能是能根据系统时间显示不同的问候语，如 Good morning(afternoon、evening、night 等)。

(1)新建 Web 项目，项目结构如图 16-3 所示。

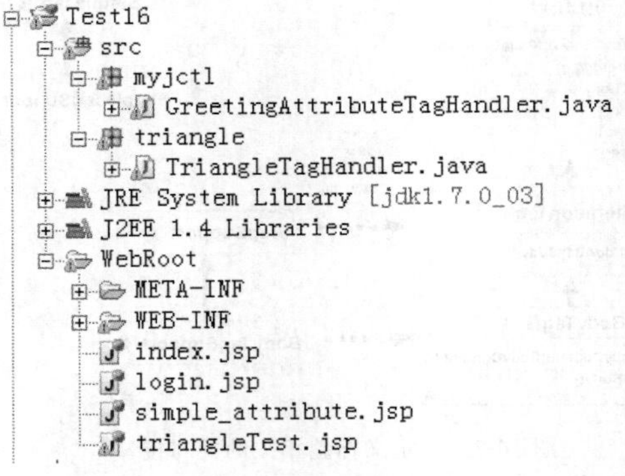

图 16-3 项目结构图

(2)创建 GreetingAttributeTagHandler 类作为标签的处理类。因为标签<simple:greeting/>是空标签，只需要实现 doEndTag()方法即可，返回 EVAL_PAGE 后继续后续页面的执行。

源代码如下：

```
//带属性的简单标签处理类
package myjctl;
import javax.servlet.jsp.*;
import javax.servlet.jsp.tagext.*;
```

```
import java.util.*;
import java.io.*;
public class GreetingAttributeTagHandler extends TagSupport{
    //标签有一个属性:name
    private String name;
    public String getName(){
        return name;
    }
    public void setName(String name){
        this.name=name;
    }
    public int doStartTag() throws JspTagException{
        Calendar calendar=Calendar.getInstance();
        int hr=calendar.get(Calendar.HOUR_OF_DAY);
        int min=calendar.get(Calendar.MINUTE);
        int sec=calendar.get(Calendar.SECOND);
        String showtime=",现在时间是 :"+ hr+ ":"+ min+ ":"+ sec;
        try{
            if(hr<12){
                pageContext.getOut().write("Good monrning,"+ getName()+showtime);
            }
            else if (hr<16){
                pageContext.getOut().write("Good afternoon,"+ getName()+showtime);
            }
            else if (hr<19){
                pageContext.getOut().write("Good evening,"+ getName()+showtime);
            }
            else{
                pageContext.getOut().write("Good night,"+ getName()+showtime);
            }
        }catch(IOException e){
            throw new JspTagException(
                    "Fatal error:greeeting tag could not write to the output stream.");
        }
        return EVAL_BODY_INCLUDE;
    }
    public int doEndTag() throws JspTagException{
        return EVAL_PAGE;
    }
}
```

(3)在如图 16-4 所示的目录中新建标签库描述(TLD)文件,命名为 greeting.tld。

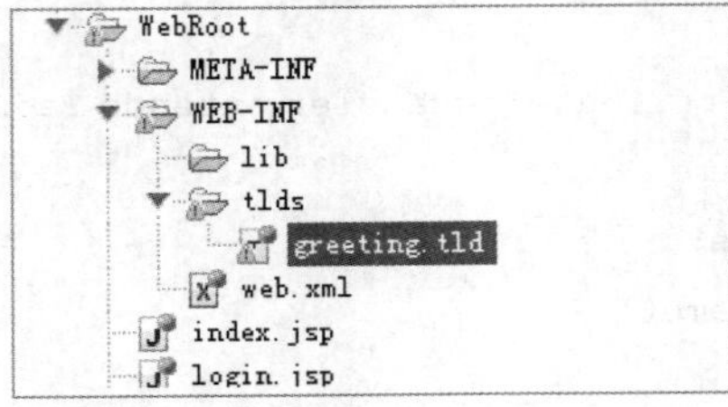

图 16-4 TLD 文件存放路径

源代码如下:

```
<?xml version="1.0" encoding="UTF-8"?>
<taglib version="2.0">
    <tlib-version>1.0</tlib-version>
    <short-name>simple</short-name>
    <tag>
        <name>greetingAtt</name>
        <tag-class>myjctl.GreetingAttributeTagHandler</tag-class>
        <body-content>JSP</body-content>
        <attribute>
            <name>name</name>
            <required>true</required>
            <rtexprvalue>true</rtexprvalue>
        </attribute>
    </tag>
</taglib>
```

(4)在 web.xml 文件中配置标签库信息。

源代码如下:

```
<?xml version="1.0" encoding="UTF-8"?>
<web-app version="2.4"
    xmlns="http://java.sun.com/xml/ns/j2ee"
    xmlns:xsi="http://www.w3.org/2001/XMLSchema-instance"
    xsi:schemaLocation="http://java.sun.com/xml/ns/j2ee
    http://java.sun.com/xml/ns/j2ee/web-app_2_4.xsd">
  <welcome-file-list>
    <welcome-file>index.jsp</welcome-file>
  </welcome-file-list>
  <jsp-config>
      <taglib>
          <taglib-uri>/CustomTags</taglib-uri>
          <taglib-location>
              /WEB-INF/tlds/greeting.tld
          </taglib-location>
      </taglib>
```

```
    </jsp-config>
</web-app>
```

(5)在项目中新建 login.jsp。

源代码如下：

```
<%@page language="java" import="java.util.* " pageEncoding="utf-8"%>
<html>
    <head><title>登录页面</title></head>
    <body>
        <form method="post" action="simple_attribute.jsp">
                <h3>登录能获取时间、日期</h3><hr>
                请输入您的姓名：
                <input type="text" name="username">
                <p>
                <input type="submit" value="登录">
            </form><hr>
    </body>
</html>
```

(6)在项目中新建 simple_attribute.jsp,应用上面自定义的 simple 标签。

源代码如下：

```
<%@page language="java" import="java.util.* " pageEncoding="utf-8"%>
<%@page contentType="text/html;charset=utf-8"%>
<%@taglib uri="/CustomTags" prefix="simple"%>
<html>
    <head><title>简单自定义标签带属性例子</title></head>
    <body>
        <%request.setCharacterEncoding("utf-8");
            String username=request.getParameter("username");%>
        <font size="4" color="blue">
        <simple:greetingAtt name="<% =username% >" />
        </font>
    </body>
</html>
```

项目完成后,部署到 WebLogic 上,运行效果如图 16-5 和图 16-6 所示。

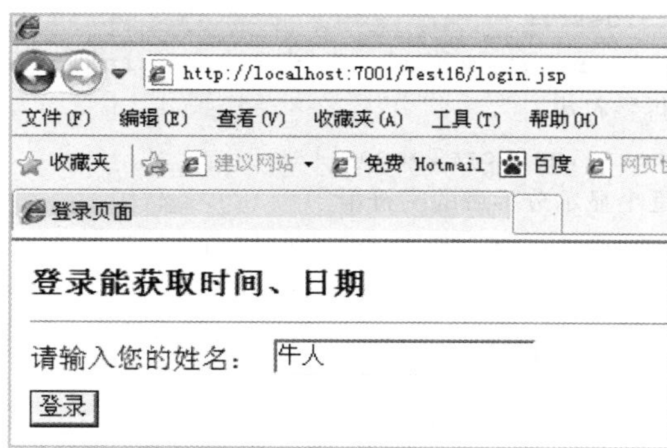

图 16-5 输入姓名

图 16-6 使用自定义的 simple 标签显示姓名和系统时间

2.创建<triangle:showMessage/>标签，该标签的功能是允许用户输入三角形的三边，如果三条边能构成三角形，则输出三角形的面积与三边的边长；若不能构成三角形，则显示提示信息。

(1)在上题项目中创建标签处理类。

```
//带属性的简单标签处理类
package triangle;
import javax.servlet.jsp.*;
import javax.servlet.jsp.tagext.*;
import java.util.*;
import java.io.*;
public class TriangleTagHandler extends TagSupport{
    //标签有一个属性:name
    private String sides;
    public String getSides(){
        return sides;
    }
    public void setSides(String sides){
        this.sides=sides;
    }
    public int doStartTag() throws JspTagException{
        int i=0;
        String s=null;
        double result=0;
        double a[]=new double[3];
        String answer=null;
        s=this.getSides();
        if(s!=null){ StringTokenizer fenxi=new StringTokenizer(s,",");
        //根据","分拆字符串
            while(fenxi.hasMoreTokens()){
                //逐个显示分拆后的字符串
                String temp=fenxi.nextToken();
                  try{
                        a[i]=Double.valueOf(temp).doubleValue();
                        i++;
                  }catch(NumberFormatException e){
```

```
                    answer="<BR>"+ "请输入数字字符";
                }
            }
            if(a[0]+a[1]>a[2]&&a[0]+a[2]>a[1]&&a[1]+a[2]>a[0]){
                double p=(a[0]+a[1]+a[2])/2;
                result=Math.sqrt(p*(p-a[0])*(p-a[1])*(p-a[2]));
                answer="面积:"+result;
            }
            else{
                answer="<BR>"+"您输入的三边不能构成一个三角形";
            }
        }
String showMsg="<P> 您输入的三边是:<BR>"+a[0]+ "<BR>"+a[1]+ "<BR>"+a[2]+
"<BR>"+ answer;
        try{
            pageContext.getOut().write(showMsg);
        }catch(IOException e){
            // TODO Auto-generated catch block
            e.printStackTrace();
        }
        return EVAL_BODY_INCLUDE;
    }
    public int doEndTag() throws JspTagException{
        return EVAL_PAGE;
    }
}
```

(2)在上题的 greeting.tld 文件中追加新标签的定义。

源代码如下:

```
<?xml version="1.0" encoding="UTF-8"?>
<taglib version="2.0">
    <tlib-version>1.0</tlib-version>
    <short-name>simple</short-name>
    <tag>
        <name>greetingAtt</name>
        <tag-class>myjctl.GreetingAttributeTagHandler</tag-class>
        <body-content>JSP</body-content>
        <attribute>
            <name>name</name>
            <required>true</required>
            <rtexprvalue>true</rtexprvalue>
        </attribute>
    </tag>
    <short-name>triangle</short-name>
```

```
    <tag>
        <name>showMessage</name>
        <tag-class>triangle.TriangleTagHandler</tag-class>
        <body-content>JSP</body-content>
        <attribute>
            <name>sides</name>
            <required>true</required>
            <rtexprvalue>true</rtexprvalue>
        </attribute>
    </tag>
</taglib>
```

(3)在 web.xml 文件中配置标签库信息。

源代码如下：

```
<?xml version="1.0" encoding="UTF-8"?>
<web-app version="2.4"
    xmlns="http://java.sun.com/xml/ns/j2ee"
    xmlns:xsi="http://www.w3.org/2001/XMLSchema-instance"
    xsi:schemaLocation="http://java.sun.com/xml/ns/j2ee
    http://java.sun.com/xml/ns/j2ee/web-app_2_4.xsd">
  <welcome-file-list>
    <welcome-file>index.jsp</welcome-file>
  </welcome-file-list>
  <jsp-config>
      <taglib>
          <taglib-uri>/CustomTags</taglib-uri>
          <taglib-location>
              /WEB-INF/tlds/greeting.tld
          </taglib-location>
      </taglib>
      <taglib>
          <taglib-uri>/TriangleTags</taglib-uri>
          <taglib-location>
              /WEB-INF/tlds/greeting.tld
          </taglib-location>
      </taglib>
   </jsp-config>
</web-app>
```

(4)新建页面 triangleTest.jsp，使用自定义的标签<triangle:showMessage>来对三角形的三边进行判断。

源代码如下：

```
<%@page contentType="text/html;charset=GB2312" %>
<%@taglib prefix="triangle" uri="/TriangleTags"%>
<HTML>
```

```
<BODY BGCOLOR=cyan><FONT Size=3>
  <P>请输入三角形的三个边的长度,输入的数字用逗号分隔:
  <BR>
  <FORM action="" method=post name=form>
    <INPUT type="text" name="sides" value="${param.sides}">
    <INPUT TYPE="submit" value="送出" name=submit>
  </FORM>
  <triangle:showMessage sides="${param.sides}"/>
</BODY>
</HTML>
```

项目完成后,部署到 WebLogic 上,运行效果如图 16-7 和图 16-8 所示。

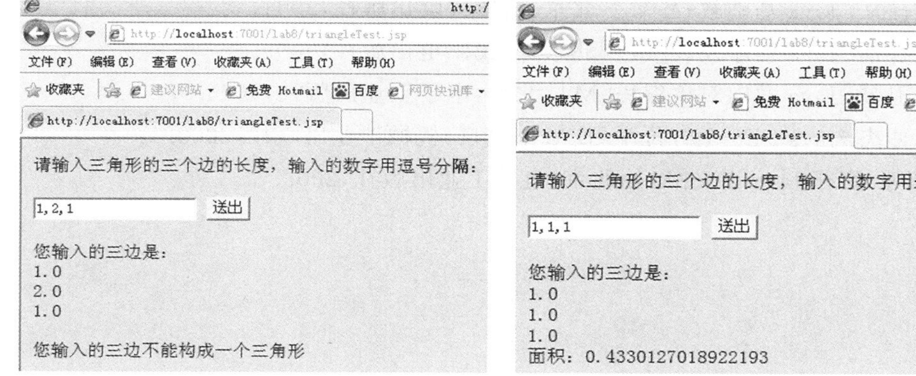

图 16-7 输入的三条边不能构成三角形

图 16-8 输入的三条边能构成三角形则计算面积

16.5 实验总结

本次实验主要练习如何使用标签文件来自定义用户标签,让读者熟悉创建自定义标签的基本步骤,体验自定义标签的使用过程。利用 JSP 自定义标签,软件开发人员和页面设计人员可以各自独立工作,页面设计人员可以把精力集中在使用标签(HTML、XML 或者 JSP)创建网站上,而软件开发人员则可以将精力集中在实现底层功能上。

16.6 课后思考题

1.若要使用自定义标签,需要调用什么指令,调用标签的格式如何书写,遇到标签的起始标记和结束标记时,分别调用什么方法?

2.解释概念:标签库、标签库描述文件、标签处理类。

3.要自定义空标签,需要继承什么类或实现哪些接口?

参考文献

[1] 张屹,吴向荣. 企业级 Java Web 编程技术——Servlet & JSP[M].大连:大连理工大学出版社,2012.

[2] IBM 培训服务部. 企业 Core Java 开发技术[M]. 天津:天津科技翻译出版公司,2010.

[3] 耿祥义. Java 基础教程[M]. 北京:清华大学出版社,2004.

[4] 朱喜福. Java 程序设计[M]. 北京:人民邮电出版社,2009.

[5] 孙卫琴. Java 面向对象编程[M]. 北京:电子工业出版社,2006.

[6] 朱仲杰. Java SE6 全方位学习[M]. 北京:机械工业出版社,2008.

[7] 李刚. 疯狂 Java 讲义[M]. 北京:电子工业出版社,2008.